W9-AYH-182

PHYSICS

A CRASH COURSE

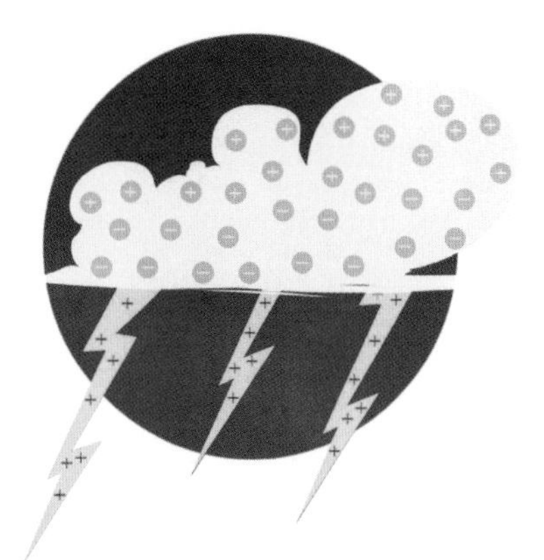

PHYSICS

A CRASH COURSE

Brian Clegg

METRO BOOKS
New York

METRO BOOKS
New York

An Imprint of Sterling Publishing
1166 Avenue of the Americas
New York, NY 10036

METRO BOOKS and the distinctive Metro Books logo
are trademarks of Sterling Publishing Co., Inc.

ISBN: 978-1-4351-6969-2

For information about custom editions, special sales,
and premium and corporate purchases, please contact
Sterling Special Sales at 800-805-5489
or specialsales@sterlingpublishing.com.

Manufactured in China

10 9 8 7 6 5 4 3 2 1

sterlingpublishing.com

Credits: Art Director James Lawrence;
Design JC Lanaway;
Illustration Nick Rowland and Andrea Ucini

Cover illustrations Andrea Ucini

INTRODUCTION

The word "physics" conjures up a wide range of responses. At school, drawing ray diagrams for light or dealing with irritating equations for forces can seem a tedious exercise with little point. Yet physics is also the most fundamental of the sciences—it tells us how the universe works. It's behind most of our exciting technology, from space rockets and satellites to cell phones, from electric cars to MRI scanners. Even better, in the twentieth century, physics underwent a revolution. Relativity and quantum theory transformed it from a bread-and-butter subject to a gourmet feast of fascinating oddities and intriguing possibilities. Whether it's the reality of time travel or quantum teleportation, modern physics teases the mind with its daring.

Founding figures

While elements of physics were known to the ancient Greeks—particularly Archimedes—and developed to some extent during the medieval period, physics in the modern sense began with Galileo Galilei in the early seventeenth century. Galileo's great contribution was *Discourses and Demonstrations Relating to Two New Sciences* (1638), which covers forces and motion, though he ventures into a number of byways, including the nature of infinity. His book gives information on how things fall under gravity, using the controlled method of pendulums and rolling balls down inclined planes. And it discusses the way that projectiles move.

What was dramatically different about Galileo's approach was that his ideas were based on experiments, both real and based in thought. There is a small amount of mathematics in Galileo's work, but it was his successor, Isaac Newton, who brought numbers to the fore. Newton, working at the end of the sixteenth century, made physics a mathematical science with his laws of motion and gravitation, and his book, the *Principia*, provided a huge step forward toward modern physics.

Always relative

As far as the laws of motion and gravitation go, Newton's work was accurate enough to take us to the Moon. It would not be until Einstein took on relativity at the start of the twentieth century that Newton would have to be fine-tuned for the unexpected relationship between time and space that becomes apparent at very high speeds, and for aspects of gravitation that emerged from Einstein's answer to Newton's biggest unresolved problem: how gravitation acted at a distance. By bringing in the concept of warps in spacetime, Einstein gave us a total picture of the weakest—but most obvious—force of nature.

Albert Einstein was the first true media scientist—so much so that we tend to associate the term "relativity" uniquely with him. However, not only did several contemporaries make key steps toward the special and general theories, the concept of relativity dates back to Galileo. It was he who pointed out that we always have to ask the question "Relative to what?" when considering motion. Galileo's relativity is about ensuring that we think about context for movement. He pointed out that aboard a windowless, smoothly traveling boat, there would be no way to tell from experiments inside that you were moving. When on the boat, it isn't in motion as far as you are concerned—it's the water and the Earth below that's moving backward.

This concept of Galilean relativity, though, is no trick with words. It has a genuine and important effect on how we interact with the world around us. It's Galilean relativity that means we aren't left behind by the moving Earth—and that increases the speed of impact when two cars are in a head-on collision. If each travels at 50 miles per hour (80 km/h), they will collide at 100 mph (160 km/h), relative to each other. Similarly, it's Galilean relativity that requires aircraft to take off into the wind. The wind speed adds to the speed of the air across the wings, enabling the plane to take off at a lower ground speed. Einstein once joked about Galilean relativity while on a train, asking, "When does the station arrive at this train?"

Expanding views

In Newton's day, physics was primarily concerned with forces and movement, with astronomy still pigeonholed in the Greek fashion under mathematics. However, as the discipline became more clearly established, not only would considerations of the universe come into the fold, but other natural phenomena became more clearly identified and added to the remit of physics. Both electricity and magnetism had been known since ancient times, but for many centuries they remained little more than a mystery. Magnets become increasingly useful in compasses, helped by the idea developed in the sixteenth century by English natural philosopher William Gilbert that the Earth itself was a huge magnet and it was this that made compass needles point in a particular direction. Gilbert constructed elegant little magnetic globes known as terrellae to investigate this. Meanwhile, by the eighteenth century, electricity had become an entertainment for the wealthy with displays such as the "Flying Boy," where a youth was suspended from silk cords, charged up with static electricity, and used to pick up lightweight objects by electrostatic attraction and to produce sparks from his fingertips.

However, it would be the nineteenth century, as with so much else in science and technology, that saw electricity and magnetism come to the fore as a topic of physics. English scientist Michael Faraday, and other pioneers of the period such as the American Joseph Henry, produced a whole range of electrical and magnetic discoveries, leading to the development of the electric generator and motor. Their work made it clear that magnetism and electricity did not exist in isolation from each other. Moving magnets produced electricity, while electric currents could be used to create electromagnets.

The observed phenomena were pulled together by the Scottish physicist James Clerk Maxwell. He not only gave a mathematical basis for the combined field of electromagnetism, but also showed that electromagnetic waves should exist—and if they did, they would have to travel at the speed of light, establishing the origin of one of the key aspects of nature.

Getting heated

At the same time, the steam engine, the motive power of the Industrial Revolution, had produced a need to know more about heat and energy. As a result of this, thermodynamics, the part of physics dealing with transfer of heat and heat engines was developed, and the different forms of energy were unified into a single concept.

This was the ideal time to be opening up thermodynamics. To get a proper picture of what was happening to, for example, the steam in an engine, it was necessary to take a statistical view of its atomic components. The concept of atoms, first developed in ancient Greek times, was revived and improved to explain how chemical reactions took place—but it also proved essential in getting an understanding of the behavior of matter as it was heated and cooled.

The detailed understanding of what was happening at the atomic level became a new interest of physicists, particularly as it was discovered that atoms had structure. With first the electron and soon after the atomic nucleus found to exist, it became clear that there was more to the nature of matter than simple spherical atoms. The study of matter at this level would lead to one of the biggest areas of physics in the twentieth century, particle physics.

Patently genius

Finally, it was time for that best-known of physicists, Albert Einstein, to crash onto the scene. In one remarkable year, 1905, Einstein published four papers, each of which was arguably worthy of a Nobel Prize. To make this feat even more remarkable, he was not yet even a working academic. At the time he was a clerk in the Swiss Patent Office.

Not only was Einstein's achievement remarkable, in those four papers, he managed to span all of the areas of major development in physics during the twentieth century. The first paper provided evidence for the existence of atoms (which was still in question in 1905) and an estimation of the size of water molecules. Next came the paper that won him the Nobel Prize in Physics for 1922—not on relativity, but on a phenomenon called the photoelectric effect. This established the existence of the photon, which helped lay the foundations for quantum theory, one of the two main planks of modern physics.

The other two notable papers *were* on relativity. One paper described the special theory of relativity, which shows how space and time are linked, and demonstrated how fast-moving objects will be influenced as their time slows down, will contract in the direction of movement, and will increase in mass. The other paper, a short addendum to the special relativity paper, produced Einstein's most famous equation (though not in this form): $E=mc^2$.

Einstein would go on in the subsequent ten years to develop his general theory of relativity, explaining why Newton's gravitational law works as a result of matter warping space and time, while the warp in spacetime tells matter how to move. The general theory also added a number of factors that made a more accurate prediction of gravitational behavior, and its equations would lead to predictions of everything from black holes to the way that the universe evolves. The picture of physics as we know it today was brought into existence in the first half of the twentieth century.

Getting on course

Our crash course in physics is divided into 52 sections and these are split into four chapters. The first takes in matter and light. The nature of "stuff" has always been of interest to philosophers and later scientists. This chapter takes on the atomic components of matter, what mass is, and how atoms are held together to form larger structures. We explore the different forms that matter can take—solids and liquids, gasses and plasma, matter's strange twin antimatter and the possibility that the universe contains far more of an invisible substance called dark matter.

Although matter is a vital part of the universe, however, just as important a constituent is light. It's light that carries the Sun's energy across space and makes the Earth warm enough for life to exist. And when we get onto quantum physics in Chapter three, we will discover that light is also the glue that holds matter together. But, for the moment, we explore the whole range of light's spectrum from radio and infrared through to X-rays and gamma rays. We look at familiar features of light, such as color and reflection and the subtler behavior of refraction and polarization. Finally, we discover light's most important quality—its remarkable speed.

Light is a form of energy, which leads us neatly into the second chapter, dealing with energy and heat. Here we see the formal definitions of energy, work, and power—and how these differ from the ways that the words are commonly used. We explore the different types of energy, among them kinetic, potential, chemical, and nuclear. Machines come in next as we look at deployment of energy, followed by the very specific form of kinetic energy that is heat, plus how heat engines make use of this before moving on to temperature. The extreme cold limit of absolute zero takes us into the final sections of this chapter on the laws of thermodynamics and the nature of entropy.

There's a distinct change of gear from the nineteenth to the twentieth century as we move into Chapter three and quantum physics. The starting point is Maxwell's equations relating to electricity and magnetism, giving us the bridge between energy and the quantum. We discover what "quantum" means, and how the young Danish physicist Niels Bohr would finally explain how the atom works. From here we bring in the central structures of quantum physics, such as Schrödinger's equation and the uncertainty principle. This leads on to some of the implications of quantum physics, how light and matter can be both particles and waves, and the extra forces necessary for the atomic nucleus to exist. After bringing in the standard model of particle physics and the essential concept of fields, we conclude with two of the later developments in quantum physics: quantum electrodynamics and entanglement.

Finally, in Chapter four, we bring in the other greater transformer of physics, relativity. To deal with this we first need to have the concepts of mechanics. We begin with the nature of empty space and movement, adding in force and acceleration to take on Newton's laws. Before plunging into relativity, we also encounter friction and fluid dynamics. Initially, relativity is very straightforward, and the work not of Einstein, but Galileo. But then we bring in the special theory of relativity, and via the concept of gravity, the general theory. Finally, we deal with one of the significant implications of the general theory—the black hole—and the way that general relativity has enabled physicists to model the whole universe.

In these 52 sections we've gone from inside an atom to the edge of the universe. What else could it be but a crash course in physics?

How to use this book

This book distills the current body of knowledge into 52 manageable chunks, allowing you to choose whether to skim-read or delve in a bit deeper. There are four chapters, each containing 13 topics, prefaced by a set of biographies of the leading physicists and timelines of key events in the history of physics. The introduction to each chapter gives an overview of some of the key concepts you might need to navigate.

Each topic has three paragraphs.

The Main Concept provides a theory overview.

PARTICLES & WAVES

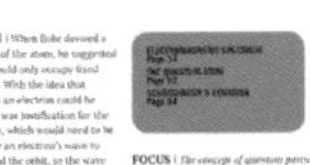

The Drill Down looks at one element of the main concept in more detail, to give another angle or to enhance understanding.

The Focus is an extra nugget of information or key memorable fact.

"Do not Bodies and Light act mutually upon one another; that is to say, Bodies upon light in emitting, reflecting, refracting and inflecting it, and Light upon Bodies for heating them, and putting their parts into a vibrating motion wherein heat consists?"

ISAAC NEWTON,
OPTICKS (1704)

1

MATTER & LIGHT

INTRODUCTION

Until the middle of the nineteenth century, physics dealt with forces and gravity, light, electricity, and magnetism. It was about the way in which objects behaved, plus the insubstantial oddities of nature. Exploring the composition of substances was left to the chemists.

However, two nineteenth-century breakthroughs would ensure that physics took on a wider remit. The first was over the behavior of atoms and molecules in gases. Scottish physicist James Clerk Maxwell and German physicist Ludwig Boltzmann independently worked on what would become known as statistical mechanics, which described the behavior of gases by treating them as a statistical collection of many interacting particles. This made it possible to use basic physics to deduce many of the physical properties of matter made up from those particles.

Matter became even more part of the essence of physics when in the twentieth century quantum physics (see chapter 3, pages 80–113) revealed the nature of atoms and their unique structure. It was realized that the way different elements formed chemical bonds was entirely dependent on the physics of the atomic structure. Even the periodic table, beloved of school chemistry labs and TV quizzes, would prove to be nothing more than a structural diagram of the layout of electrons in the outer layers of the atoms in question.

Heart of the matter

At its heart, physics is all about establishing the basics, the foundations. This is why the nature of matter has become a significant aspect of physics, while the details of the behavior of different chemicals is left to the chemists and biologists. An old joke, beloved of physicists emphasizes their focus on the fundamentals, which can lead to a surprising lack of detail.

A geneticist, a nutritionist, and a physicist are arguing about the best way to produce the perfect racehorse. The geneticist says: "Well, of course, it's a matter of breeding stock. Getting the best line genetically, breeding the right animals with a perfect pedigree and getting the ideal outcome." The nutritionist replies: "Of course I accept the importance of genetics, but in the end, it's what you feed the horse that makes the difference between a winner and a loser." The physicist smiles, shakes her head, turns to her whiteboard and starts to write an equation. "Let's assume the racehorse is a sphere," she says.

The components and makeup of matter, then, have become significant parts of the physics story. However, there is a reason why matter is linked together in this chapter with what appears to be a very different thing: light. We think of matter as far more tangible. Gases may seem at first glance to be similarly insubstantial as a beam of light—until we think of the impact of a hurricane, nothing more than moving gas. Crucially, all matter has mass, where light has none. However, since Einstein's work, which we'll come back to in more detail in the fourth chapter, the division between matter and light has become fuzzy at best. We know that matter (m) can be transformed into energy (E) and vice versa according to the relationship of $E=mc^2$—and the form of energy involved in this transformation is light (c).

Linking light and matter

Although all of physics in some way influences our everyday lives, matter and light make up all of our experience. Our bodies and all the objects around us are made up of matter. Knowing more about the nature of the atoms within that matter, and how they are connected with bonds gives us a better picture of how everything around us functions.

Matter may seem to come in many different forms. And it's been estimated that there are around 10^{80} atoms of matter in the universe—that's 1 followed by 80 zeroes. Yet all that matter is made up of arrangements of atoms of around 100 different elements, themselves all formed from a handful of fundamental particles.

Light is also far more than "the thing that allows us to see." All light produced after the big bang originates in matter. It is when matter loses energy, notably when an electron drops down to a lower energy level around an atom, that a photon of light is produced. In addition an invisible form of light acts as the carrier of the electromagnetic force, which means that pieces of matter don't pass straight through each other. Light and matter are inextricably intertwined.

Physics is concerned at the fundamental level with what makes up matter, the different forms that matter can take, such as solids, liquids, and gases, and how the atoms within matter interact. When studying light, physics gives us insights into how light gets from place to place and how it interacts with matter. The late Stephen Hawking suggested you should "Look up at the stars and not down at your feet"—but in reality, physics encourages us to do both.

BIOGRAPHIES

JOHN DALTON (1766–1844)

Born in the north of England in 1766, John Dalton had a deep interest in the sciences but was barred from a traditional academic career as he was a Quaker. He picked up information where he could and at the age of 27 began to teach mathematics and science in Manchester. This lasted seven years until the college he worked for got into financial difficulties. Dalton became a private tutor, which remained his main source of income until receiving a government pension.

Dalton was active in the Manchester Literary and Philosophical Society, which encouraged him to explore subjects as wide as the nature of color blindness (he was color blind himself), the weather, and the behavior of gases. This last study inspired Dalton to think about the way that substances combined to make chemical compounds.

Over a number of years up to 1803, Dalton put together a theory of atomic behavior. Various elements made of tiny components called atoms, either singly or in combination, linked up to make matter. Dalton gave each element a relative weight; his values were often incorrect, but his vision was remarkable, given that his equipment was crude even by the standards of the time.

GILBERT LEWIS (1875–1946)

In the physics community, Gilbert Lewis is best remembered for naming the photon. But this pioneering American chemist had a huge impact on the study of the structure of matter. Lewis studied at Harvard before settling into academic life at the University of California, Berkeley.

A versatile thinker, Lewis made considerable advances in the energy changes in chemical reactions and on the nature of acids and bases. He extended the definition of an acid, adding the "Lewis acids." He was also the first to produce heavy water, where the hydrogen in H_2O is replaced with deuterium, a heavier (and rare) isotope or version of hydrogen that has a neutron as well as a proton in its nucleus. He even contributed to relativity theory. But his most important discovery involved bonding.

Lewis came up with the idea that atoms could link up when their outer electrons were shared to form a connection that was later termed a covalent bond. He never won the Nobel Prize despite being nominated 41 times. Lewis died in his laboratory while working on cyanide. His death certificate gives the cause of death as heart disease, though strong cyanide fumes were found in the laboratory.

PAUL DIRAC (1902–1984)

Paul Adrien Maurice Dirac, born in Bristol, UK, gained two degrees at Bristol University before moving to Cambridge for a PhD. His most significant contribution to physics would be helping to develop quantum electrodynamics—the theory of the interaction of light and matter—and producing the Dirac equation.

One aspect of Dirac's equation was to extend Schrödinger's equation describing the evolution of a quantum system to include the effects of relativity. However, the most surprising impact of this work was the prediction that there was a different kind of matter—antimatter—in the form of a negatively charged equivalent of an electron. This particle, known as the positron, was discovered soon after.

Dirac became Lucasian Professor of Mathematics at Cambridge, and like an earlier holder of this position, Isaac Newton, had very limited social skills. Famously, while taking questions after a lecture, an audience member said that he didn't understand something Dirac had written. When asked why he didn't respond to this, Dirac said that it was a statement, not a question. Dirac won the Nobel Prize in Physics in 1933, along with Erwin Schrödinger.

VERA RUBIN (1928–2016)

American astronomer Vera Rubin was born Vera Cooper in Philadelphia in 1928. Already engaged enough to build her own telescope as a child, she went on to study astronomy at Vassar College. After a brief spell at Cornell, her main academic institutions were Georgetown University and the Carnegie Institute, while her main work was on the rotation of galaxies.

After studying a number of galaxies, including our neighbor Andromeda, Rubin discovered that the outer regions of a galaxy spun around as fast as the more central stars. That was not expected. One possible reason for this was that there was far more mass in the outer parts of the galaxies than was thought. What's more, taking spiral galaxies as a whole, many seemed to be rotating so fast they should be flying apart. Rubin's observations would give strong support to the concept of dark matter, an unknown form of matter that only interacted through gravity. This had been first postulated by Swiss astronomer Fritz Zwicky in the 1930s, but largely ignored.

Rubin won many honors for her work but never the Nobel Prize. She died in 2016 and is celebrated as someone who helped lead the way for women in astronomy and physics.

5th century BCE

GREEK MODEL
Ancient Greek philosopher Leucippus and his pupil Democritus propose a theory of matter where each substance was composed of small particles known as atoms. The theory is rejected by Aristotle as it required empty space between the atoms, which he believes could not exist. Aristotle's views would hold sway for many centuries.

1803 CE

DALTON'S THEORY
English natural philosopher John Dalton devises a new atomic theory, reflecting the experiments on gases and elements that had been undertaken in the previous 20 years. He suggests each element is made from spherical atoms with a specific weight, while compounds contain fixed proportions of elements.

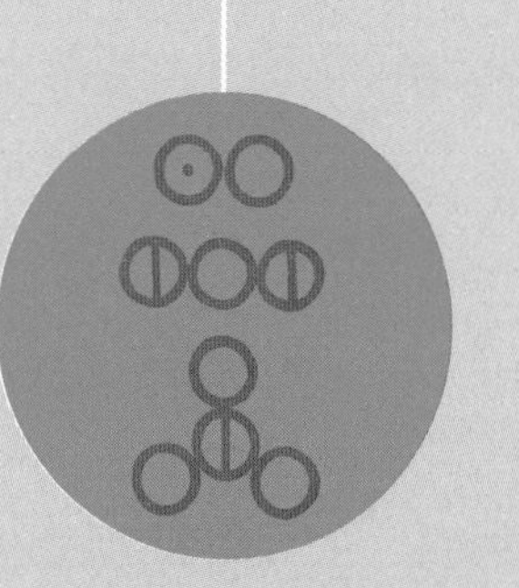

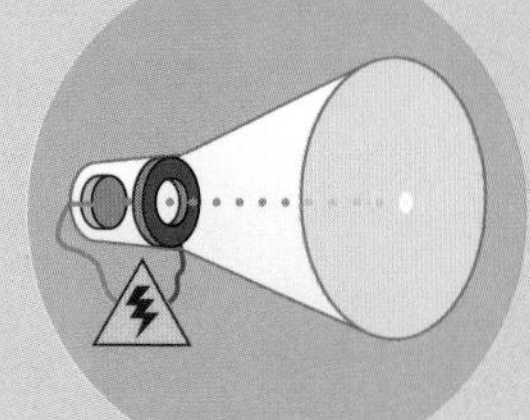

1897

ELECTRONS
English physicist J. J. Thomson realizes that cathode rays, produced by electrical voltages across near-vacuums in glass tubes, are streams of electrically charged particles, later called electrons, which have significantly smaller mass than could be the case for any atom. This is the first evidence that, should atoms exist, they are not indivisible spheres as Dalton thought.

BROWNIAN MOTION
German physicist Albert Einstein, working as a patent clerk in Bern, Switzerland, works out a mathematical model of Brownian motion. This phenomenon involves tiny particles (such as soot or fungal spores) suspended in water, which dance around as if alive. Einstein shows their behavior is consistent with bombardment by moving water molecules, giving both an idea of the size of the molecules and direct evidence of their existence.

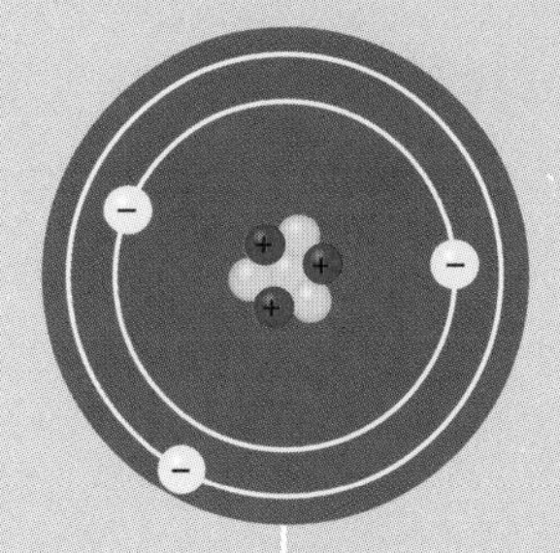

BOHR'S DISCOVERY
Danish physicist Niels Bohr completes the model of the hydrogen atom by finding a way that electrons could exist in the space around the nucleus without being attracted inward and plunging into the core of the atom. His approach depends on the new concept of quantum mechanics.

1909

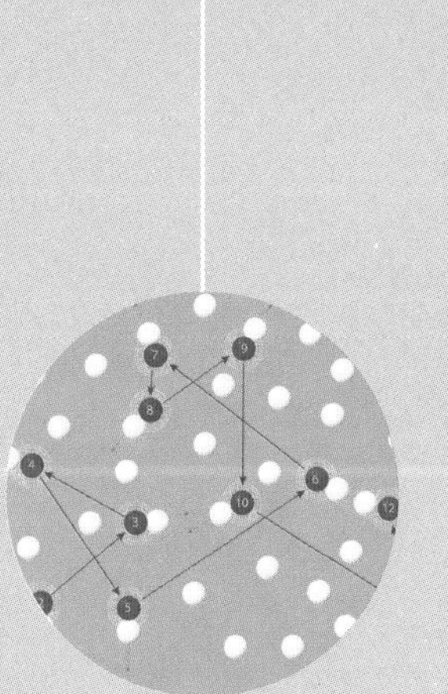

NUCLEAR ATOM
New Zealand physicist Ernest Rutherford, working in Manchester, England, discovers that atoms are not uniform spheres with electrons distributed through them, as Thomson thought, but had a small, very dense, positively charged nucleus at their heart and electrons somewhere around the outside of the atom.

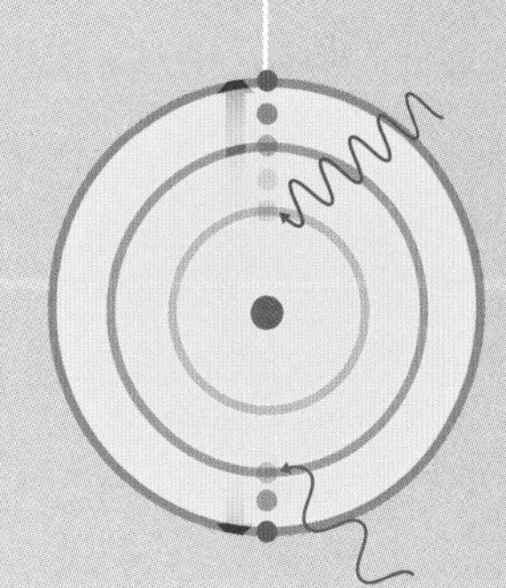

ATOMS

THE MAIN CONCEPT | By the nineteenth century the idea that matter consisted of the four elements of earth, water, air, and fire, which dated back to the ancient Greeks, was seeming untenable. Experiments on different gases showed that, for example, the air contained a number of different substances such as oxygen, nitrogen, and carbon dioxide. To replace the four elements, the outmoded concept of atoms from the Greek *a-tomos*, meaning "uncuttable," was revived, initially by the English natural philosopher John Dalton. It took a good 100 years before atoms were entirely accepted as being anything more than a useful tool for calculation, but now we know that there are around 94 types of atom in nature (with a few more artificial ones), which make up all the matter in the universe. Although the original concept of the atom was that it was the smallest possible unit of a substance, around the start of the twentieth century it became clear that atoms themselves had a structure of smaller components. The central nucleus consisted of positively charged protons and electrically neutral neutrons, while around the outside of the atom were the much lighter negatively charged electrons. Each type of atom is the basis for one of the chemical elements. Elements are substances that cannot be broken down into simpler constituents. Atoms of each elements have a unique number of subatomic particles that sets them apart from the rest.

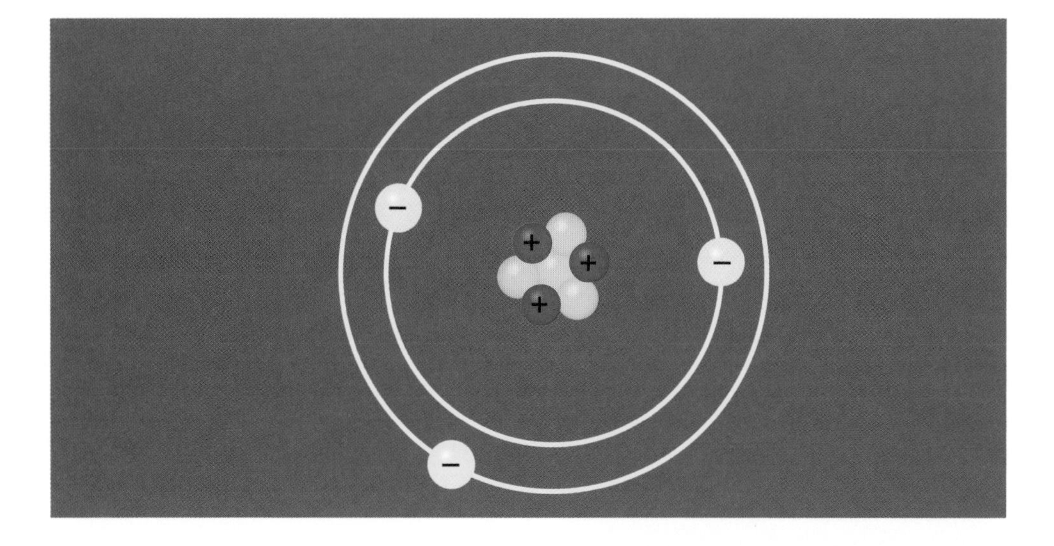

DRILL DOWN | The number of atoms in nature is often given as 92, running from the lightest, hydrogen, to the heaviest, uranium. Most of the heavier atoms are unstable, decaying to produce lighter-in-weight elements. However, plutonium, atomic number 94, has been discovered in space, so it is now more appropriate as the heaviest natural atom. All of the lightest element, hydrogen, most helium, and a small amount of lithium are thought to have been made in the big bang. All the other atoms up to iron (number 26) were made in stars by the fusion process that powers them. Still heavier elements need more energy and were produced in stellar collisions and exploding stars known as supernovae.

MASS
Page 22
BONDING
Page 24
THE QUANTUM ATOM
Page 92

FOCUS | *Physicists tend to think of their discipline as more fundamental than chemistry or biology, but when New Zealand-born physicist Ernest Rutherford needed a name for the dense central core of an atom, he looked for inspiration to biology. The cells that make up multicelled organisms had a central body, called a nucleus. Rutherford named the atomic nucleus after this.*

MASS

THE MAIN CONCEPT | The term "mass" is easier to get a feel for than accurately describe. Until Newton's day, there was little distinction between mass and weight—and in everyday usage we still don't differentiate them, yet they are two quite different things. Newton, who was the first to use the word "mass" in this context, defines it as "a measure of matter that arises from its density and volume jointly." Some have argued that this definition is circular, as density is usually defined in terms of mass. However, in essence, mass defines the amount of "stuff" in an object. The mass of an object doesn't change whether it's on Earth, on the Moon, or floating in space—the amount of stuff present remains the same. Weight, by contrast, is dependent on location because it is the force of gravity acting on a body, so is dependent on both the mass and the acceleration due to gravity where the body is located. In principle, an object could have two different masses—its gravitational mass that produces its weight, and its inertial mass, which shows how much it resists acceleration when it is pushed by another force. However, both these masses have always appeared to be identical in size.

ATOMS
Page 20
MOMENTUM & INERTIA
Page 126
GENERAL THEORY OF RELATIVITY
Page 142

DRILL DOWN | The assumption that inertial and gravitational mass are identical lies behind Einstein's masterpiece, the general theory of relativity. He claimed to have had his "happiest thought" sitting in a chair in the Swiss Patent Office, thinking, "If a person falls freely he will not feel his own weight." This is true. It's why astronauts float about on the International Space Station. The strength of gravity there is around 90 percent of that on Earth, but in orbit, the space station is in free fall toward the ground—it just happens to be moving sideways as well, so continually misses the Earth. Einstein's statement implies that inertial and gravitational mass are identical.

FOCUS | *The kilogram is the SI (International System of Units) unit of mass. If my mass is 70 kg, then my weight is 686.7 newtons (the scientific unit of force). On the Moon it would be 113.8 newtons. We often describe weight using kilograms, assuming a pull from Earth's gravity. By contrast, the pound is a US customary system unit of weight. The equivalent mass unit is the slug, although it is rarely used and mass is often also measured in pounds.*

BONDING

THE MAIN CONCEPT | Atoms alone are not enough to explain the nature of matter. Because each atom contains electrical charges—a positively charged central nucleus surrounded by negatively charged electrons—two or more atoms can be attracted together electrically, a process known as bonding. On a small scale, this bonding produces molecules, which are units of more than one atom. These can be formed from identical atoms—a molecule of oxygen has two oxygen atoms in it—or different atoms—a molecule of water has one oxygen and two hydrogen atoms. Molecular bonds are either "covalent," where electrons are shared between the bonded atoms, or "ionic." In this latter case, a positively charged ion (an atom that has lost one or more electrons) and a negatively charged ion (an atom that has gained electrons) are attracted together. In a gas, molecules act independently, but in a liquid, weaker bonds caused by attraction between the nucleus of one atom and the electrons of another, mean that the atoms stick together loosely. In a solid, where atoms are not jiggling around so vigorously, the bonds are stronger, forming large-scale structures where every atom is linked to its neighbors. Some solids, known as crystals, have links between atoms that form a regular, repeating lattice structure; other solids are amorphous and without a regular pattern.

DRILL DOWN | The degree to which elements can form bonds is dependent on the electronic structure of the atom. Electrons around the atom inhabit concentric "shells," each with a maximum capacity. The innermost shell can take two electrons, the next eight, the third 18, and so on. The number of electrons in the shell of number n equals the formula $2(n^2)$. For example, shell four has a capacity for 32 electrons (2×4^2), and shell seven—the outer limit of the largest atoms discovered so far—can hold a whopping 98 (2×7^2). Bonds tend to be made which fill the outer shell. So, elements where the outer shell is already filled—the noble gases, such as helium and neon—rarely form bonds. The reason carbon is so versatile and ubiquitous in life is that with four electrons and four gaps in its outer shell, its atoms are very flexible in the structures they can make.

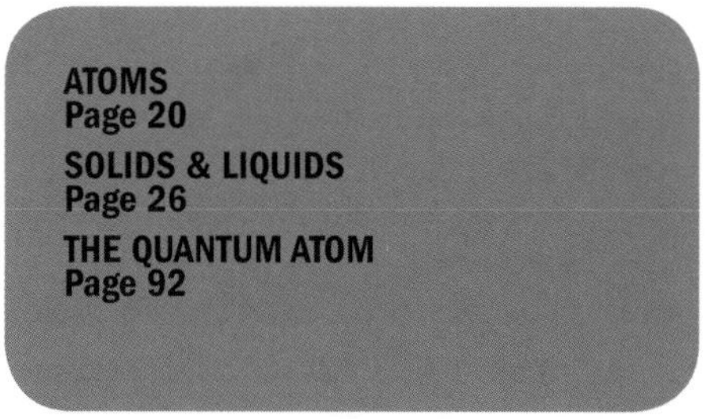
ATOMS
Page 20
SOLIDS & LIQUIDS
Page 26
THE QUANTUM ATOM
Page 92

FOCUS | *It is the "hydrogen bonding" in water that makes it possible for water to be liquid on Earth. In water, the positive charge of the hydrogen's nucleus is attracted to the oxygen's negative electrons, meaning it takes extra energy to get the molecules to separate as a gas. Without hydrogen bonding, water would boil at around –94°F (–70°C), making the Earth uninhabitable.*

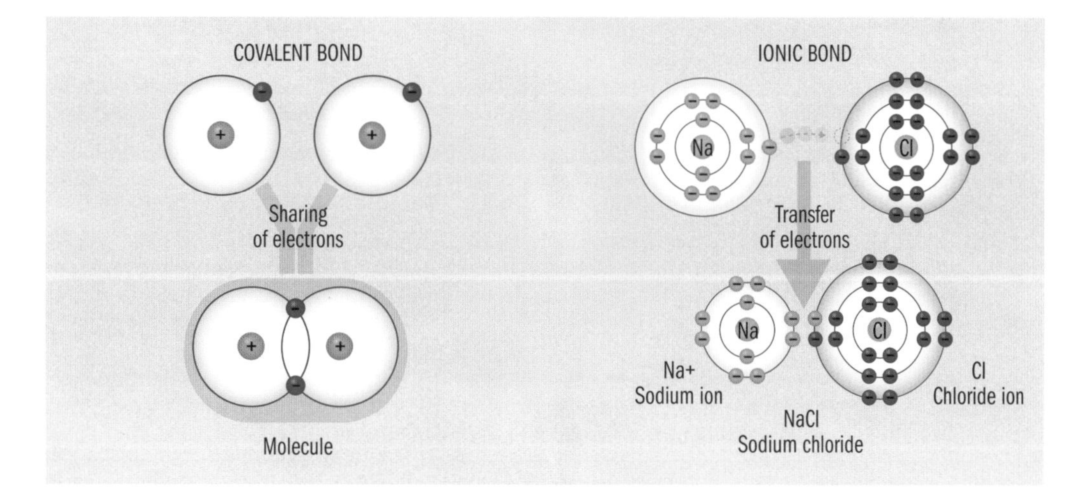

SOLIDS & LIQUIDS

THE MAIN CONCEPT | Solids and liquids are the states of matter we most directly experience. Solids are usually the densest state that matter can be in, although a few solids, such as ice, are less dense than their densest liquid form, as the shape of the crystals leave sufficiently large gaps to reduce density. Because of the rigid bonds between atoms, solids have distinct shapes that they tend to keep despite forces applied to them, although their response to excess force can be anything from flexing to shattering and crumbling. Liquids have relatively weak bonding, taking the shape of their containers under gravity or a sphere in zero gravity. The properties of solids tend to depend on their atomic structures and the way the atoms are fitted together in the solid. For example, in some solids, a number of the electrons are able to move relatively freely through a crystal lattice, making the solid a conductor, while in others the electrons are firmly tied to the atoms and the result is an insulator. Liquids can also be electrical conductors. Here, the current is passed by ions that move through the liquid. For example, pure water is a good insulator, but it usually contains ions that make it able to conduct relatively well.

ATOMS
Page 20
BONDING
Page 24
ELECTROMAGNETISM
Page 88

DRILL DOWN | The borderline between solid and liquid is not always as obvious as it might appear it should be from the simple description. Some liquids have such a strong attraction between molecules that they become very viscous, most dramatically in bitumen (pitch). The Brisbane pitch drop experiment, running since 1927, features pitch in a funnel. So far only nine drops of pitch have fallen from it. Other, so-called non-Newtonian liquids, change viscosity under stress. For example, a suspension of corn starch thickens under pressure, where a non-drip paint thins. Gels come closest to being a hybrid. They are flexible solids which are mostly liquid but have enough structure to retain integrity.

FOCUS | *Glass is sometimes mistakenly identified as a very viscous liquid, but it is an amorphous solid. The misunderstanding may have arisen from ancient window panes, which tend to be thicker on the bottom edge. Once thought to be caused by the liquid running very slowly, it is because glass was rarely even in those days and the thickest edge would be put at the bottom for stability.*

GASES & PLASMA

THE MAIN CONCEPT | Gases and plasmas are both loose collections of matter. In a gas the atoms or molecules are in their normal state, while a plasma is much hotter (the Sun is a ball of plasma) and so it is made up of ions—atoms or molecules that have an electric charge because they have either gained or lost electrons. Unlike liquids and solids, the particles in gases and plasmas are moving sufficiently quickly that there is very little influence from the attraction between them, so they will spread out to fill a container entirely. The constant battering of the particles against a container's walls collectively provides the pressure of the gas or plasma. As temperature is a measure of the speed of the particles, pressure will increase with temperature—a feature of gases known as Gay-Lussac or Amontons' law. Similarly, at constant pressure, volume changes with temperature (Charles' law), while pressure changes inversely with volume at constant temperature (Boyle's law)—the gas pressure rises as you squeeze it into a smaller volume. While a gas is not usually a conductor, a plasma's electrically charged atoms or molecules conduct well. A lightning bolt passes through air because the extremely high voltage ionizes the gases in the air. Because plasmas are electrically charged, they can be contained or shaped without walls using electromagnetic fields.

DRILL DOWN | Although in principle we could work out how a gas or plasma behaves by tracking every atom or molecule, in practice this isn't possible. In the nineteenth century, statistical mechanics was first used to predict the behavior of gases. Ideas such as temperature and pressure are not absolutes but the result of statistical combination of the action of many particles. Gas molecules move quickly—in air at room temperature they go about 1,640 feet (500 meters) per second. But because they constantly collide, they don't get far before being deflected. As a result, molecules get from place to place at a relatively slow pace, which is why smells take time to diffuse through the air.

FOCUS | *We tend to think of plasma as something unusual and dramatic—but the vast majority of matter in the universe is in the plasma state. This is because stars are mostly plasma and dominate solar systems—over 99 percent of the mass in our solar system is in the Sun—and there is highly diffuse plasma in the space between galaxies.*

ATOMS
Page 20

MOMENTUM & INERTIA
Page 126

FLUID DYNAMICS
Page 134

ANTIMATTER

THE MAIN CONCEPT | Antimatter sounds fictional but it does exist. Every type of matter particle has an antimatter equivalent, which has inverted values of some of its properties. Where the particle is electrically charged, the antiparticle has the opposite charge. So, for instance, the positron (the antimatter equivalent to an electron) is positively charged, while the antiproton is negatively charged. If the particle isn't charged, it still has an antiparticle with other properties, such as the orientation of its magnetism, reversed. The positron was predicted from theory by Paul Dirac in 1928 and detected in 1932 by Carl Anderson. Antiparticles are much less abundant than normal matter. They are produced in small amounts by nuclear reactions as seen in radioactivity. However, we rarely detect them because when an antimatter particle meets the equivalent matter particle the pair annihilate into energy in the form of photons. Conversely, energy can produce matter and antimatter pairs. According to the big bang theory, the very early universe consisted solely of energy, but this converted into matter and antimatter. That proposes that there are equal amounts of matter and antimatter in the universe, leaving it as something of a mystery as to where all the antimatter is. Positron-emitting materials, generating high energy photons, are used in PET (Positron Emission Tomography) medical scanners, while at the CERN laboratory near Geneva, antihydrogen has been produced—atoms with antiproton nuclei and positrons replacing electrons.

ATOMS
Page 20
NUCLEAR ENERGY
Page 64
THE STANDARD MODEL
Page 106

DRILL DOWN | The problem of the missing antimatter in the universe has puzzled physicists ever since the big bang theory was proposed. There is no evidence of large quantities of antimatter in nearby galaxies—we would expect to see dramatic annihilation on its borders with matter. It is, though, possible that distant galaxies are made of antimatter. There would be no obvious visible difference, but they would be detectable if a matter galaxy collided with them. It seems more likely that some kind of asymmetry in nature, possible with current theories of particle physics, meant that slightly more matter than antimatter was produced. Once all the antimatter had been annihilated, the remaining matter is what makes up our universe.

FOCUS | *Although antimatter would in principle make a great fuel as matter/antimatter annihilation packs in 1,000 times as much energy as nuclear fuels, it is also the most expensive substance in the world. In its simplest form as positrons, it has been estimated to cost around $25 billion for a gram—and at the moment, annual production is only measured in millionths of a gram.*

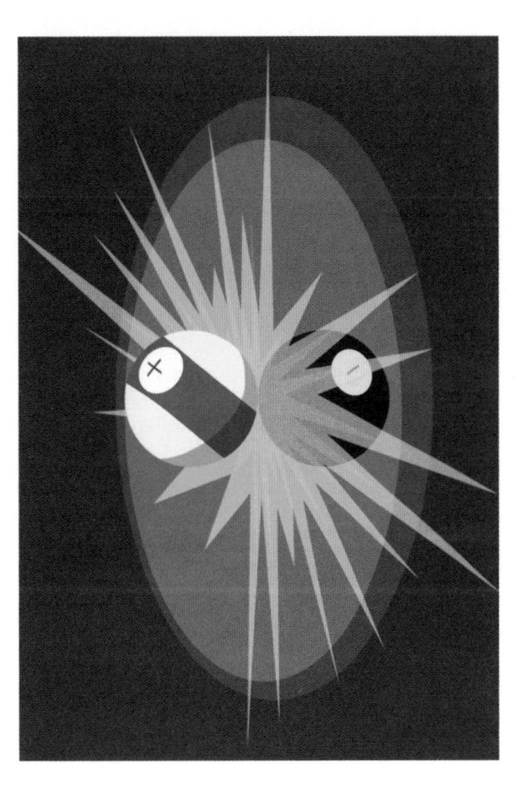

DARK MATTER

THE MAIN CONCEPT | Since the 1930s, there have been concerns that galaxies and clusters of galaxies rotate too quickly. Think of a lump of clay on a potter's wheel. If the clay is spun too fast, pieces fly off. Similarly, if a galaxy spins too quickly, the gravitational force between the stars won't keep them in place. Many galaxies and clusters of galaxies appear to rotate too quickly for gravity to hold them together, given the matter we can detect in them. The best supported explanation is that the galaxies contain large quantities of an unknown matter particle that cannot be seen or detected other than by its gravitational pull. This is dark matter, a translation of "dunkle Materie," named by Swiss astronomer Fritz Zwicky and given more detail by American astronomer Vera Rubin. Although the presence of dark matter in a galaxy does explain its ability to rotate so quickly, it doesn't yet explain all gravitational anomalies and dark matter particles have never been detected. One alternative possibility is that the equations of gravity need to be altered when dealing with such vast bodies. The best-known approach to this, modified Newtonian dynamics, explains most observed behavior, though it isn't yet a perfect solution. Another possibility is that the calculation of the mass of galaxies is inaccurate.

DRILL DOWN | Most existing theories on the nature of dark matter assume that it is made up of vast quantities of a single type of particle. However, this assumption is increasingly being questioned. After all, we know that ordinary matter is not based on a single fundamental particle. In total the standard model of particle physics includes 12 different particles that could be considered matter particles. It makes little sense to assume that dark matter is far less complex than the matter we can observe. Because no dark matter has ever been directly discovered, the jury is still out on its composition—or whether it exists at all.

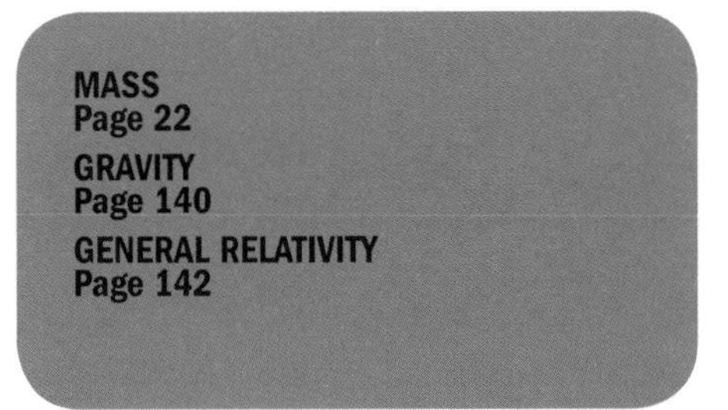

MASS
Page 22
GRAVITY
Page 140
GENERAL RELATIVITY
Page 142

FOCUS | *The most startling thing about dark matter is that (assuming it exists) there is so much of it. It is estimated that there is over five times as much dark matter by mass as there is ordinary matter in the universe. For this calculation, even the most exotic known matter—black holes, for example—is included in the "ordinary matter" figure.*

ELECTROMAGNETIC SPECTRUM

THE MAIN CONCEPT | If we consider the "stuff" that makes up the universe, light is just as significant as matter because energy in the form of light has proved to be interchangeable with matter. For centuries, the nature of light was debated, but by the nineteenth century it was accepted that it traveled as a wave (the picture would later be complicated by quantum theory). When the Scottish physicist James Clerk Maxwell was developing his theory of electricity and magnetism in the 1860s, he realized that interacting waves of electricity and magnetism could be self-sustaining and would travel at around the speed of light. By this time both infrared and ultraviolet forms of nonvisible light had been discovered—Maxwell's electromagnetic waves allowed for a whole spectrum far beyond the visible. Longer wavelengths would be discovered in the 1880s with the addition of radio waves, followed in the next decade by short wavelength X-rays and gamma rays. (Although gamma rays are generally shorter wavelength than X-rays, the distinction between the two is usually the method of production. X-rays are produced when high-energy electrons hit atoms, while gamma rays come from the atomic nucleus.) The shorter the wavelength of the waves, the more energy they carry, making X-rays and gamma rays particularly dangerous.

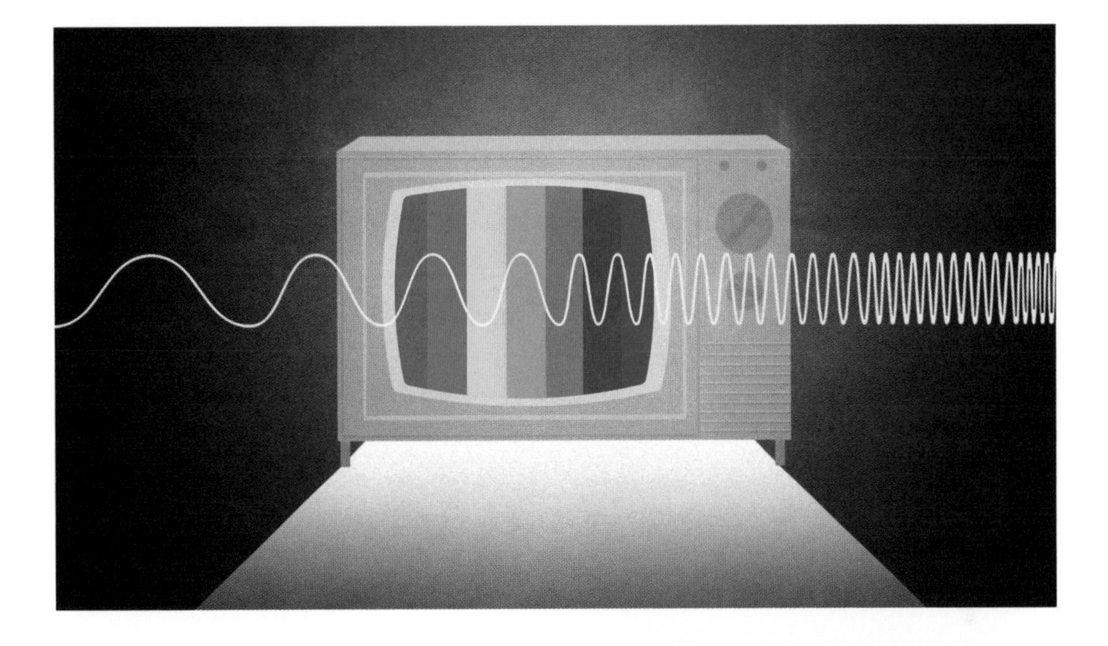

DRILL DOWN | Maxwell showed that electromagnetic waves were self-sustaining. A varying electric field produced a varying magnetic field, which produce a varying electric field and so on, traveling as a pair of waves at right angles to each other. Previously, it had been assumed that there must be a medium that filled all space for light waves to travel through. This was obvious with sound waves, where the medium was air—remove the air and we no longer hear sound. But light crossed empty space, so an invisible material called ether was devised to provide the necessary medium. All attempts to detect the ether failed, and with Maxwell's theory it no longer proved necessary.

COLOR
Page 36

QUANTA
Page 90

PARTICLES & WAVES
Page 102

FOCUS | *In 1800, astronomer William Herschel was experimenting on light at his home in Slough, England. He used a prism to produce a spectrum, selected a thin slice of color and let it fall on a thermometer. The temperature increased as he moved from blue to red. But it continued to go up well after the visible spectrum ended. He had accidentally discovered infrared radiation.*

COLOR

THE MAIN CONCEPT | Color is an interpretation of the wavelengths of light being seen. However, our perception of color is made more complex by the nature of color vision. Our eyes do not have a single type of color detector, but three different ones, which are most sensitive to blue, green, and yellow/red. The perception of color depends on the combinations of these three, allowing us to see colors that aren't on the spectrum. For example, there is no magenta in the spectrum—it's what we see when light has much of the rest of the spectrum but no green. Sunlight is pretty much white (we only think of the Sun as yellow because a lot of its blue light is scattered in the sky, making it look more yellow), but this is made up of all the colors of the spectrum. When white light hits an object, that object will typically re-emit (or reflect) some but not all of the wavelengths. What we see as the object's color are the wavelengths the object doesn't absorb well. So, for example, a red apple is absorbing most of the colors in the white light and only re-emitting red. The primary light colors are red, green, and blue, while pigments that absorb these appear to have opposing colors: cyan, magenta, and yellow.

DRILL DOWN | Isaac Newton was the first to produce a detailed description of light and color. In a now famous experiment he allowed a small beam of light from a hole in his wall to fall on a prism, producing a spectrum. He then took a small segment of the spectrum and sent it through a second prism. At the time, it was assumed that the glass was producing the colors. If so, the second prism should change the color of the segment again—but it didn't. Newton had demonstrated that the white light contained the different spectral colors. He confirmed this by recombining the colors to white with a lens.

ELECTROMAGNETIC SPECTRUM
Page 34
REFLECTION & REFRACTION
Page 38
PARTICLES & WAVES
Page 102

FOCUS | *We can all list seven colors of the rainbow—red, orange, yellow, green, blue, indigo, and violet—but it's a lot harder to see them in a spectrum. In reality, there are many different colors, but the eye cannot distinguish them. It was Newton who decided on the traditional seven, probably to match the musical notes A to G.*

REFLECTION & REFRACTION

THE MAIN CONCEPT | Reflection of light off a surface is the simplest optical process. Without reflection, we wouldn't be able to see objects, while reflection in shiny surfaces, such as polished metal, produces the familiar mirror image. In classical physics, reflection is assumed to be a process where light bounces off a surface, like a ball hitting a wall. The angle at which the light rays hit the surface is equal to the angle at which they reflect. But this is not a very satisfactory picture. Light can't bounce. Instead, the more modern quantum picture shows us that reflection involves atoms in the reflecting surface absorbing incoming light, then re-emitting it. Something more complex happens when light hits the surface of transparent material—it refracts. In ordinary materials, the light rays bend inward toward the perpendicular as they move to a denser material, for example from air to glass. This effect is caused because light travels slower in denser materials such as water and glass than it does in air. Refraction is why a straw appears to bend at the point it enters the drink, and a glass of water appears to reverse the view behind it. Different wavelengths of light are refracted by different amounts, which is how rainbows form inside raindrops and prisms produce a spectrum.

DRILL DOWN | The degree to which light bends in refraction is determined by the refractive index of the two mediums. This is the ratio of the speed of light in a vacuum to its speed in a medium. The refractive index of glass, for example, is around 1.5, while air is a fraction above 1. However, in recent years special materials known as metamaterials have been developed, which have negative refractive indexes—the light bends the opposite way to expectations. This effect produces remarkable results. For example, microscopes with conventional lenses (which bend light due to refraction) can only see down to around the wavelength of the light used, but with metamaterial lenses it is possible to resolve much smaller objects.

FOCUS | *A familiar reflection effect is mirror reversal. Your image appears to have left and right side swapped. But why does a mirror reverse left and right but not top and bottom? In reality, it swaps front and back (hold a book up and its front cover becomes the back). We simply assume that the rotated left hand is the image's right hand.*

COLOR
Page 36

SPEED OF LIGHT
Page 44

PARTICLES & WAVES
Page 102

PRINCIPLE OF LEAST ACTION/TIME

THE MAIN CONCEPT | A fundamental principle of nature is the principle of least action, and the associated principle of least time. The principle says that nature is lazy; a moving body takes the route involving least effort. For example, if you throw a ball both forward and up, it will take the path where its average kinetic energy minus its average potential energy is least. If you make this assumption, it's possible to derive the laws of motion. In the seventeenth century, French mathematician Pierre de Fermat came up with a related statement, the principle of least time, sometimes called Fermat's principle. He said that the path a ray of light will take between points is the one that will take the light the least amount of time to traverse. So, for example, if light goes from air, into water where it travels slower, it will follow a path that allows it to spend more time in the air and less in the water than if it continued on a straight line. This is a quantum effect—in motion, quantum particles like photons take every possible path, with different probabilities, each represented by a wave of probability. The waves for most paths cancel out, while the route with least time is reinforced. This is exactly what happens in refraction.

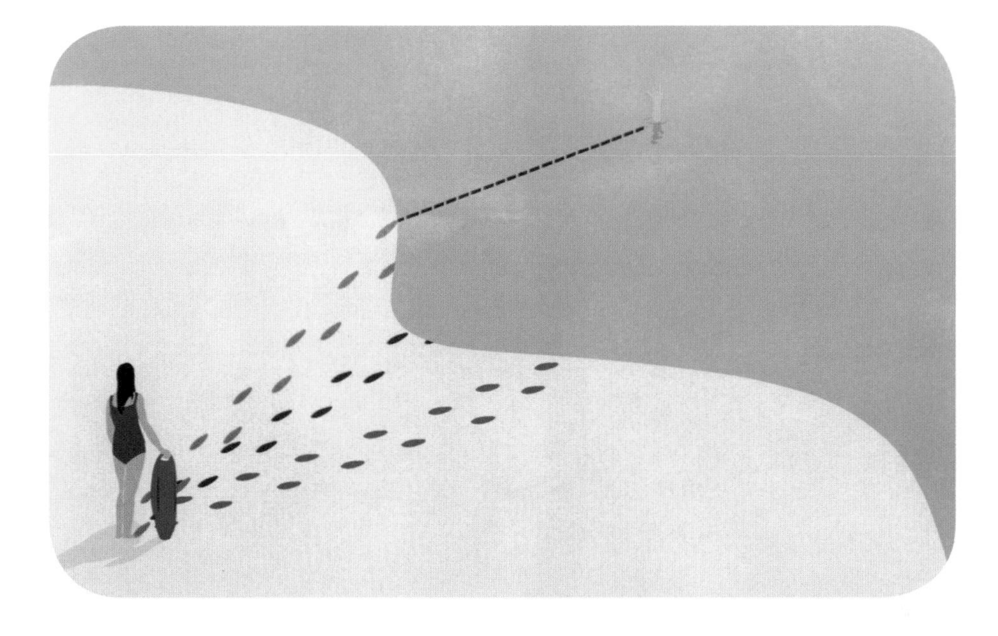

DRILL DOWN | The principle of least time is sometimes called the Baywatch principle. This is because it is put to practical use by beach lifeguards. The lifeguards may be excellent swimmers, but they can still run much faster than they can swim. If a lifeguard sees someone struggling in the water, the natural instinct may be to run straight toward them. However, the better approach is to run toward a point on the edge of the sea that is closer to the same position left-to-right on the beach as the swimmer is in the sea. That way, the lifeguard goes a longer distance on the beach, and a shorter distance through the water.

REFLECTION & REFRACTION
Page 38
QED
Page 110
NEWTON'S LAWS
Page 130

FOCUS | *The American physicist Richard Feynman used the principle of least action in his PhD thesis, looking at the way that a quantum particle could get from A to B. Instead of assuming that it would travel in a straight line, he applied probabilities to each of the possible routes the particle could take, which proved highly effective in explaining quantum behavior.*

POLARIZATION

THE MAIN CONCEPT | Light waves consist of two waves, electric and magnetic, both at right angles to each other, and always at right angles to the direction of motion. The polarization of the light concerns the orientation of the plane (or flat space) in which the electric wave is moving. In ordinary light, waves are vibrating in a random set of planes all orientated in different directions. However, some interactions of light, such as reflection, tend to select specific polarization directions, in which case the light has become "polarized." This is why Polaroid sunglasses, which filter out horizontal polarization, reduce glare. Some materials, notably the form of calcite known as Iceland spar, refract light to two different angles depending on its polarization, producing two images of anything seen through it. This has been used in aircraft bomb sites to estimate distances. By far the biggest modern application of polarization is in LCD (liquid-crystal display) screens. In the screen, a liquid crystal is sandwiched between two polarizing filters at right angles to each other. Initially the filters prevent any light getting through. But when an electric current is passed through segments of the screen, the crystal twists the polarized light, allowing it through the second filter.

DRILL DOWN | There are two types of polarization. In simple "linear" polarization caused, for example, by reflection or a polarizing filter, the direction of polarization remains the same as the light progresses. However, it is also possible to have circular polarization, where the direction of polarization rotates with time. As the light moves forward, the polarization spirals around its direction of travel. Circular polarization can be set up in either direction, making it usable for applications such as separating left and right eye images in 3D movies. Most commonly circular polarization is achieved using special filters, but it is occasionally found in nature, most notably in reflection from the carapace of some beetles.

ELECTROMAGNETIC SPECTRUM
Page 34
ELECTROMAGNETISM
Page 88
PARTICLES & WAVES
Page 102

FOCUS | *Harvard student Edwin Land was fascinated by polarized light. In 1926, aged 18, he took leave of absence from university and developed a material where very small polarizing crystals were embedded in a sheet of plastic. The crystals were oriented to all exclude horizontally polarized light. When this "Polaroid" material was oriented correctly it cut out reflected glare and began Land's Polaroid company.*

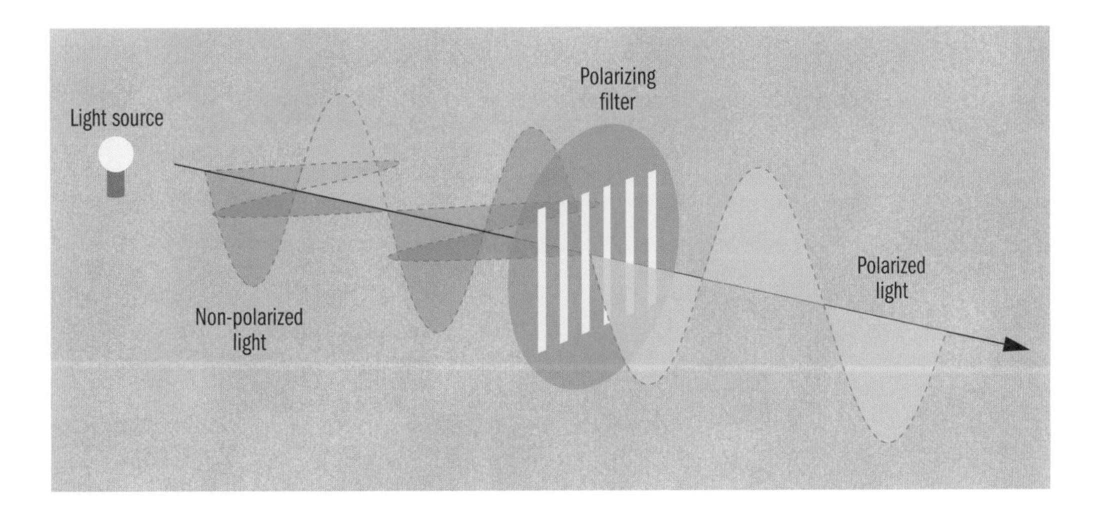

SPEED OF LIGHT

THE MAIN CONCEPT | For many years it was debated whether light traveled infinitely fast or just extremely quickly. The first to have evidence that there was a finite speed of light was Danish astronomer Ole Rømer in 1676. Rømer was hoping to use the moons of Jupiter as a clock to help with navigation at sea, but he discovered something strange. Rather than appearing to move at a constant rate, the moons seemed to slow down for a period of time, then began to speed up. The timing seemed linked to the relative positions of Earth and Jupiter. Rømer realized that his measurements were being influenced by the time it took light to reach Earth. The speed suggested by Rømer's measurements was about 220,000 kilometers per second (137,000 miles per second)—not a bad first attempt at the actual figure of approximately 300,000 kilometers per second (186,000 miles per second). This approach required accurate knowledge of the distance to Jupiter's moons, but in the nineteenth century, Armand Fizeau and a little later Jean Foucault would use fast-rotating mechanical devices to pin down the speed to around 298,000 kilometers per second (185,000 miles per second). Unlike most constants of nature, we now know the speed of light exactly—because we use it to define the unit of distance. Light speed is 299,792,458 meters per second (186,282 miles per second), because a meter is defined as 1/299,792,458th of the distance light travels in a second.

ELECTROMAGNETIC SPECTRUM
Page 34
ELECTROMAGNETISM
Page 88
SPECIAL RELATIVITY
Page 138

DRILL DOWN | Light's finite speed has a fundamental impact on astronomy. The idea that the farther something is away, the longer the light takes to get to us means that when we look out into space, we also look back in time. Light from the Sun takes a little over eight minutes to reach us—so we see the Sun as it was eight minutes ago. The next nearest star, Proxima Centauri, appears as it was around four years ago. And we see our nearest galactic neighbor, Andromeda, as it was 2.5 million years ago. The oldest light detectable is the cosmic microwave background, which has traveled more than 13 billion years.

FOCUS | *Galileo attempted to measure the speed of light. He sent an assistant with a lantern several miles up a hill at night. Galileo flashed a lantern at the assistant, who then responded. Galileo timed from sending his flash to seeing his assistant's lantern. But the time was the same as when the assistant was just feet away—he had only measured their reaction time.*

“If your theory is found to be against the second law of thermodynamics I can give you no hope; there is nothing for it but to collapse in the deepest humiliation.”

ARTHUR EDDINGTON,
THE NATURE OF THE PHYSICAL WORLD (1928)

2

ENERGY & HEAT

INTRODUCTION

Energy is one of those slippery concepts that everyone knows about but finds it very hard to describe or tightly define.

A favorite topic for spam emails is to offer free electricity if the recipient only buys the plans for a self-generating electricity device, usually, we are told, the work of the inventor Nikola Tesla (1856–1943) and made available despite attempts by the big energy companies to suppress it. Such a machine is part of a class of devices known as perpetual motion machines. These incredible—indeed, impossible—inventions are supposed to give out more energy than we put in, and provide a useful insight into the nature of energy.

For a machine to run forever with no external source of power it has to be "generating" free energy, as no real machine can be 100 percent efficient. For example, it will always waste some energy on heat through friction and air resistance. Unfortunately for the many inventors of perpetual motion machines, their devices break a fundamental law of physics—the energy in a system is conserved. It can't be made or destroyed, just transferred into different forms.

Zero-point energy

Traditionally, such inventors have ignored the problem, or claimed that conservation of energy doesn't really happen. But since the development of quantum physics, the preferred explanation has been that the machines use "zero-point energy." The good news is that zero-point energy does exist. The uncertainty principle of quantum physics tells us that, over very short periods of time, the energy of empty space can vary considerably. So, on average, even empty space has energy—this is zero-point energy.

So, it's suggested by the sellers of these devices, they are somehow able to tap into this zero-point energy in order to drive their devices. Simple. It's not at all clear how the magnets and coils in their machines provide access to this quantum energy, but let's not worry about that. Because even if there were a suitable mechanism, it would be unusable.

By definition, zero-point energy is the lowest possible level of energy in a system. And that makes it impractical to use it. Understanding why this is the case teaches us a fundamental lesson about what energy is and how we can use it.

Potential energy

The problem with harnessing zero-point energy is that to make use of energy—to get it to work—we need to be able to reach a lower level of energy. That's how energy makes things happen—by decreasing. Let's think of a few examples. Imagine being on the top of a mountain with a big rock. That rock is said to have "potential energy" because if I let it go, gravity will pull it down the mountain. As it accelerates it will gain kinetic energy of motion, which means it can hit something and do (admittedly destructive) work. A more positive equivalent would be water running down the mountainside, which can turn a water wheel or turbine and do work.

However, let's change the picture slightly. We have the same mountain top. Our rock has the same amount of potential energy. But this time, the location is surrounded by a large plateau. Let go of the rock—and nothing happens. We can't get any work out of the potential energy because there is no way of getting to a place with lower potential energy.

The same applies to a coiled spring—another form of potential energy. It can only do work if it can uncoil to a position with lower potential energy. If you constrain the spring so it can't uncoil—or if it has run down and simply can't uncoil any further—it won't do work. Or think of a battery, a chemical energy source. It can only do something if it's got a charge. If it can't go to a lower state of charge—a lower state of energy—then it's useless. It's a flat battery.

Understanding energy

So, when we return to our free electricity device, we have the same problem. Even if it could tap into the zero-point energy of space, there is nowhere lower than zero-point. It's like trying to use a flat battery or an empty fuel tank. There is no lower level to get to. Nothing can be done with it.

Doing work involves moving or converting energy, and a full understanding of what was happening was not possible until it was realized that heat was just another form of energy. Until the nineteenth century, heat was thought to be a kind of fluid that flowed from hot to cold bodies. But the need to understand steam engines better would drive a revolution in the science of thermodynamics, which concerns the movement of heat. Energy became one of the most significant aspects of physics.

BIOGRAPHIES

LORD KELVIN (WILLIAM THOMSON) (1824–1907)

Scottish physicist William Thomson was both an inventor and scientist, and as such he was as happy working on the challenge of getting a cable across the Atlantic as theoretical physics. Born in Belfast in 1824, Kelvin spent five years at Cambridge, but aged only 22 he became professor of natural philosophy at Glasgow University, remaining there throughout his career.

Kelvin's greatest work as a physicist was in thermodynamics. Although originally developed for improving steam engines, this field proved a fundamental part of physics. Kelvin helped establish the first and second laws of thermodynamics, which describe the conservation of energy and what happens as energy is moved from place to place.

The name Kelvin, which Thomson took when he became a peer in 1892, has become familiar as a standard scientific unit. The kelvin (K) is the unit of temperature on the absolute scale—this was in recognition of Kelvin's early calculation of a value for absolute zero. Although Kelvin did not contribute as much to theory as his contemporary James Clerk Maxwell, he played a significant role in establishing physics as a discipline.

JAMES JOULE (1818–1889)

It's hard to say if the English scientist James Prescott Joule, born in Salford, England in 1818, was a brewer with an interest in physics or a physicist who happened to own a brewery. Joule never attended university but had tutoring from John Dalton. His interest in energy developed from the decision of whether to use steam engines or electric motors in the brewery. In the process he discovered the relationship between electric current, resistance, and heat output.

His research on heat contributed to the overthrow of the "caloric" theory which considered heat to be an invisible fluid flowing from hot to cold bodies and was the beginning of the understanding of heat as just another aspect of energy. Joule approached this through the "mechanical equivalent of heat," showing that the amount of heat produced was identical to the work done to produce it.

Joule also collaborated with Kelvin on the absolute temperature scale. Like Kelvin, he had a scientific unit named after him—the joule (J) is now the unit of energy. Joule died in Sale in 1889 at the age of 70. Although the original Joule's brewery was demolished in the 1970s, a new brewery using the brand was opened in 2010.

SADI CARNOT (1796–1832)

French physicist Sadi Carnot was one of the early contributors to the concept of thermodynamics that lies at the heart of the physics of energy and heat. He was born in Paris in 1796. After studying at the military École Polytechnique in Paris, he served in the army for four years before taking a position with the military general staff, which meant that he was on call and free to pursue his own interests.

At the time, the capabilities of the steam engine were just being realized, and Carnot worked on the theory of these engines and how they could be improved for several years before publishing his book *Reflections on the Motive Power of Fire* in 1824. Up until then improvements to steam engines had been ad hoc. Carnot was able to abstract the basic theory behind the steam engine and show that the efficiency of a simplified engine (known as the Carnot cycle) depended on the difference of temperature between the heat of the boiler and the cold reservoir of the condenser.

Carnot left the army in 1828 and was sent to an asylum in 1832. He died from cholera in Paris that same year, aged just 36.

LUDWIG BOLTZMANN (1844–1906)

Ludwig Boltzmann would, with James Clerk Maxwell, be responsible for statistical mechanics, which describes the behavior of matter, particularly gases, from a statistical analysis of the behavior of the atoms that make it up. Born in Vienna in 1844, Boltzmann studied at the University of Vienna and he would teach at a number of German-speaking universities.

At the time, relatively few scientists accepted the existence of atoms, but Boltzmann managed mathematically to show how the behavior of a collection of atoms or molecules, interacting with each other and their surroundings, could produce the behavior we see in gases. As part of this work, he developed a new interpretation of the second law of thermodynamics. This shows that heat always moves from hotter to cooler bodies. Seen statistically, it implies that the amount of disorder in the atoms—known as entropy—always increases. Boltzmann developed a simple relationship between the entropy and the number of possible states a collection of atoms (or other components of a system) could be in.

Boltzmann committed suicide by hanging in 1906 at the age of 62. His formula for entropy was inscribed on his grave marker.

BASIC MACHINES
Ancient Greek philosopher Archimedes explains the working of basic machines, which enable energy to be transferred while multiplying force, and designs new types of machine, including the block and tackle pulley.

1712 CE

STEAM ENGINES
English engineer Thomas Newcomen invents a practical atmospheric steam engine, primarily used for pumping in mines. Later that century, in 1781, Scottish engineer James Watt patents the first practical rotary steam engine, making mobile steam engines feasible.

1800

VOLTA BATTERY
Italian scientist Alessandro Volta invents the electric cell. These devices, often linked together as a "battery" of cells, make it possible to move from studying static electricity to the current electricity that will be used in all electrical applications.

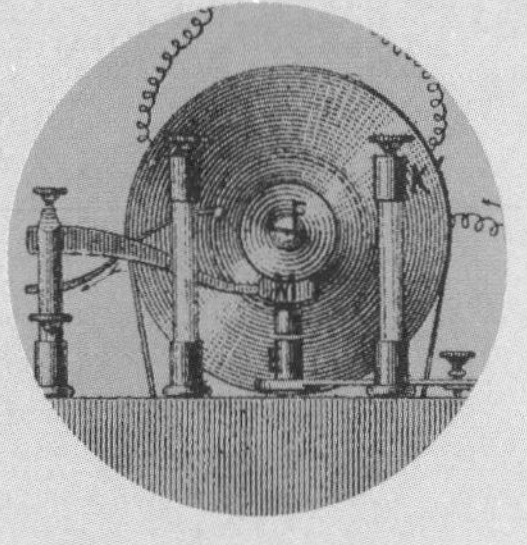

ELECTRIC MOTOR
After pulling together the existing work on electricity and magnetism, English scientist Michael Faraday makes a simple electric motor, to be followed in 1831 with his construction of the first electric generator, making the practical use of electricity possible.

1864

LIGHT WAVES

Scottish physicist James Clerk Maxwell presents his equations for electromagnetism to the Royal Society. These describe all the interactions of electricity and magnetism, not only providing the theory behind Faraday's work but also showing that light is an electromagnetic wave.

1876

AUTO POWER

German engineer Nikolaus Otto constructs the first effective gasoline engine. With German engineer Rudolf Diesel's improved diesel engine, patented in 1895, these internal combustion engines would provide the essential power source of twentieth-century transport.

1878

WATER POWER

English industrialist William Armstrong introduces the first hydroelectric power plant on his Cragside estate. Six years later, Irish engineer Charles Parsons invents the steam turbine, which would revolutionize electricity generation, used in power stations ever since.

1942

NUCLEAR POWER

A team led by Italian physicist Enrico Fermi constructs the first self-sustaining nuclear reactor in Chicago. Although the original aim was primarily to produce material for the atomic bomb program, this would be the starting point of energy generation from nuclear reactions.

WORK & ENERGY

THE MAIN CONCEPT | Energy is a natural phenomenon that produces change. It can be accumulated and stored in different ways, then used to make something happen. Energy is not made, but converted from one form to another. It comes in a whole range of forms. Potential energy is stored up somewhere, often by physical effort, to then be released. So, for instance, a rock or water at the top of a hill has potential energy, transferred as it rolls down the hill. Similarly, a wound-up spring stores potential energy. Energy is also stored in atomic bonds—this type of potential energy is referred to as chemical energy—and in the bonds of the atomic nucleus. Another group of energy types comes under the heading of kinetic energy. Where potential energy is about storage in a fixed location, kinetic energy is the energy in movement. We experience this most obviously when a fast-moving object hits us and transfers that energy to us. However, heat is also a type of kinetic energy, involving as it does the movement of the atoms making up matter. Electrical energy is combination of the potential (in, say, a battery) and the kinetic energy of the electrons moving as a current, while there is also electromagnetic energy, most familiar in the form of light. In physics the term "work" refers simply to energy being transferred from one location to another.

BONDING
Page 24
HEAT
Page 68
SPECIAL RELATIVITY
Page 138

DRILL DOWN | We speak of "losing energy" but in reality, we mean energy being transferred from place to place, or converted to a different form. The first law of thermodynamics states that energy in a closed system is conserved. The universe is the ultimate closed system because no energy can come in or leak out, but physicists define themselves all kinds of other closed systems to better understand particular phenomena. It isn't always obvious that energy is conserved because we might not notice, for example, kinetic energy being converted to heat due to friction. In the twentieth century, Einstein showed that matter and energy were inter-convertible, so strictly mass/energy is conserved. Perhaps the trickiest aspect of energy conservation is the "closed system"—when we gain energy it is from an external source, for example when the Earth is heated by the Sun.

FOCUS | *Although scientists now measure energy in units of joules, for a long time the most common scientific unit was the calorie. To add confusion, the unit often called a Calorie on food packaging (with a capital C) is actually 1,000 calories—a kilocalorie. One calorie is 4.184 joules. Meanwhile, our energy bills come in kilowatt hours, each being 3.6 million joules.*

POWER

THE MAIN CONCEPT | Although we sometimes use the term "power" correctly in everyday use—perhaps referring to the power of a car's engine—it's often used as a synonym for energy. We speak of someone having power, meaning that they are able to make things happen. In physics, though, power is the amount of energy transferred per second—the rate at which work is done. If 1 joule is transferred each second, the power is 1 watt (W), named for the Scottish engineer James Watt. When power companies use the term "kilowatt hour," they are clumsily referring to an amount of energy as "energy divided by time times time." When not considering the electrical "power supply," we often refer to power in the context of mechanical work. Here, energy used can be measured as force applied times the distance moved by an object when that force is applied to it. So, power becomes force times distance per second—or force times velocity. When considering engines, rather than use watts, we tend to revert to a much older unit of power, the horsepower. This is around 0.75 kilowatts. The output of car is usually given in brake horsepower. This is a nominal output of the engine before that power is transmitted through the mechanical parts that actually move the vehicle.

DRILL DOWN | Both the amount of energy stored in a fuel (or power source) and the power output of that fuel are significant. The amount of energy per kilogram of fuel—its energy density—tells us, for example, that kerosene is better at providing energy to a plane than lithium ion batteries, as it has 100 times the energy density. To replace 1 ton of kerosene would take 100 tons of batteries. But power output is important when we consider how the fuel is used. Gasoline has 15 times the energy density of TNT. However, TNT releases its energy much quicker. As a result of which, it makes a better explosive.

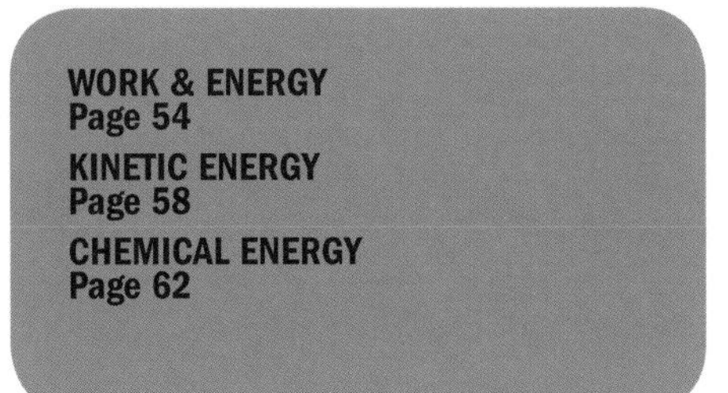

FOCUS | *It's fitting that the modern unit of power is named after Watt, because he devised the unit of horsepower. This was at a time when steam was beginning to take over from horses and Watt wanted a simple means of comparison. To make the comparison fair, one horsepower is comparable to the sustained ability of the horse, not its maximum output.*

KINETIC ENERGY

THE MAIN CONCEPT | We get the word "kinetic" from the Greek for "moving"—this is the energy of movement. Kinetic energy is needed to get something moving, and is the energy transferred when we get in the way of a moving object. It is important in understanding everything from the movement of atoms in a gas to whole galaxies. The kinetic energy of an object works out as half the mass of the object times the square of its velocity—$\frac{1}{2} mv^2$. (Velocity is speed in a particular direction.) Because the velocity is squared, an increase in speed has a major influence on the amount of kinetic energy in a moving body. This is why reducing car speeds from 30 mph to 20 in built-up areas has a significant effect. The kinetic energy difference between 30 and 20 is a ratio of 9:4. In other words, a car's kinetic energy more than doubles when increasing speed from 20 to 30 mph. To give a moving body kinetic energy, we need to take energy from elsewhere. It could be chemical energy in a bullet's explosive, a car's fuel, or your body throwing a ball. Or it could be potential energy as a rock rolls downhill. When an object slows down, kinetic energy is most often converted to heat, for example in the brakes of a car.

DRILL DOWN | A rollercoaster provides an ideal example of the way that potential energy can be converted into kinetic energy, and vice versa. The car is pulled to the top of a high slope, where it has a lot of potential energy, but has very little kinetic energy. Once the car is released, it rolls down the first incline, converting potential energy into kinetic energy as it goes. The car accelerates under gravity, its kinetic energy increasing with the square of its velocity. Kinetic energy peaks at the bottom of the slope: once the car starts upward, kinetic energy is converted back to potential energy and the car slows down.

GASES & PLASMA
Page 28

POTENTIAL ENERGY
Page 60

CHEMICAL ENERGY
Page 62

FOCUS | *When two objects collide head on and stop, it may seem that their kinetic energy has disappeared, but energy is conserved; it has to go somewhere. If the objects survive intact, all the energy will end up as sound and heat. But when, say, two cars collide, they are designed to crumple, absorbing a proportion of the energy to deform the metal structures—and divert the energy away from the passengers inside.*

POTENTIAL ENERGY

THE MAIN CONCEPT | Where kinetic energy is all about movement, potential energy is energy that is stored in the location of an object or within the structure of a substance. The most frequently used example of potential energy is due to gravity—when an object is lifted off the ground or taken to a high place. By returning to a lower point, the object can convert its potential energy into kinetic energy. However, there are far more ways to make use of potential energy, often in the bonds between atoms or molecules or bonds within the atomic nucleus. Strictly speaking, chemical energy is a kind of potential energy where the energy in those bonds can be released when the bonds are broken. But this kind of bond-based potential energy also applies, for example, to the tension in a spring. When a coil spring is wound up or a conventional spring is tensed, the bonds in the metal are stretched. The potential energy is converted to kinetic energy (and heat) when the spring is released, and the bonds pull back together. A pendulum is different from a spring in that it alternates between potential and kinetic energy as the bob moves from its highest point to its lowest and back up again.

DRILL DOWN | A dramatically powerful form of potential energy resides in the nucleus of atoms. The protons in the nucleus are all positively charged and repel each other. Getting protons close enough together to form a nucleus takes a considerable amount of energy, which goes into the strong nuclear force bonds between the protons. If those bonds can be broken, causing the nucleus to split, or undergo fission, the energy that was required to hold the nucleus together is released as heat and light. This is the source of conventional nuclear energy used in both the reactors of nuclear power stations and in nuclear fission weapons.

BONDING
Page 24

ANTIMATTER
Page 30

KINETIC ENERGY
Page 58

FOCUS | *Unlike springs, elastic bands don't store potential energy by stretching molecular bonds. Rubber is made up of long molecules which are naturally full of kinks. Stretching the band partly straightens out the kinks. But the molecules are always jiggling around due to their thermal energy and collisions between molecules push the kinks back in place, pulling the band shorter.*

CHEMICAL ENERGY

THE MAIN CONCEPT | Chemical energy, produced by breaking electromagnetic bonds between atoms to release their stored energy, was the earliest natural source of energy tapped by humans. The discovery of how to make fire, a chemical reaction that converts bond energy into heat and light, transformed human existence, not just in the ability to keep warm at night and in winter, but also in cooking, which made food safer and better able to deliver nutrition. Food is the body's source of chemical energy, which powers our existence through the process of respiration. Since the adoption of fire, chemical energy has played a role in transportation, electricity generation, and warfare. The heat released by the breakdown of bonds in long-chain carbon molecules has become ubiquitous, first from coal in steam engines and electricity generation, then from petroleum oil extracts such as gasoline, diesel, and kerosene to power vehicles, and from natural gas for home heating and electricity generation. In warfare, gunpowder and more dramatic explosives take the high-energy density and fast release of some carbon compound bonds to propel bullets and shells. Increasingly we are moving away from fossil fuels, meaning a decreasing dependence on chemical energy outside the human body, but the chances are that chemical energy will help support human civilization for decades to come.

DRILL DOWN | Although the practical applications of chemical energy are centered on the release of energy, many chemical reactions require energy to take place, storing it away in bonds. When the energy released is greater than the energy stored, the reaction gives out energy. So, for example, when hydrocarbon molecules from oil are burned, bonds between carbon and hydrogen are broken, and bonds are formed between oxygen and carbon and between oxygen and hydrogen to make CO_2 and water molecules. It takes energy to make these new bonds, but not as much as is released from the carbon–hydrogen bonds. The net result is that the amount of stored chemical energy goes down as the remaining energy is transformed into heat.

FOCUS | *We are all walking chemical energy factories. During the process of respiration, chemical constituents from our food are broken down in the presence of oxygen to release their chemical energy. Rather than generate heat, as happens in fire, cellular structures called mitochondria (which were probably once free-living bacteria) capture the energy and store it in the bonds of adenosine triphosphate for later use by the cell.*

ATOMS
Page 20

BONDING
Page 24

POTENTIAL ENERGY
Page 60

NUCLEAR ENERGY

THE MAIN CONCEPT | Just as the bonds between atoms in molecules store potential energy, so do the bonds between the components of the atomic nucleus. Depending on the elements involved, it is possible to release energy by breaking these bonds (known as nuclear fission) or by forming new, less energetic bonds (a process called nuclear fusion). Fission is used in the current crop of Earth-based nuclear power plants. Although fission can occur spontaneously with very heavy atoms, the fission process is usually initiated when a heavy atomic nucleus absorbs a neutron, producing an even more unstable nucleus. This then splits into two parts, and typically also gives off more neutrons. The energy lost from the bonds is released as heat (kinetic energy of the products) and high-energy radiation, such as gamma rays. To make use of fission in nuclear reactors (and in bombs), this process is accelerated by using the neutrons from the fission of one nucleus to stimulate the splitting of others, producing a chain reaction. Nuclear fusion, by contrast, the energy source of the Sun, is much harder to initiate. This happens when the atomic nuclei of light elements, such as hydrogen, are fused together. In the process, energy is released because there is less energy in the single final structure than in the two original nuclei.

DRILL DOWN | Nuclear fusion has the potential to be a major power source of the future. In a sense it already is, as the nuclear fusion of the Sun indirectly provides most of our energy. But for decades, work has gone into producing fusion-based generators. These would use readily available variants of hydrogen as a fuel, would not produce dangerous waste products, and would have no equivalent of a fission reactor's potential to melt down. Unfortunately, it is extremely difficult to keep a fusion reaction running, but the next generation of fusion reactors, including US-based laser fusion devices and the European ITER reactor, are expected to come close to making fusion generators practical.

FOCUS | *Austrian physicist Lise Meitner, had worked on attempts to create ultra-heavy elements by bombarding atomic nuclei. When German scientist Otto Hahn succeeded in creating nuclear fission this way, Meitner, assisted by her nephew Otto Frisch, put together the first theory of how nuclear fission could take place. By this time (1939), Meitner, who was Jewish, had fled Germany and established a new career in Sweden.*

ATOMS
Page 20

NUCLEAR FORCES
Page 104

SPECIAL RELATIVITY
Page 138

MACHINES

THE MAIN CONCEPT | A machine is a device that uses power to make something happen. The simplest machines have few parts, but still achieve this goal. Archimedes identified three—the lever, the pulley, and the screw, while by the 1500s, the wedge, inclined plane, and wheel were added to the list. Since then, the term has been applied to increasingly complicated mechanical devices, and by analogy to products such as computers (for example, the largest US computer industry body is the Association for Computing Machinery). Machines are also common in nature. Sometimes, the role of the machine is to redirect forces—this clearly happens in many of the simple machines. Other machines change energy from one form to another. Think of a power station that takes heat, uses a turbine to convert this into rotational kinetic energy, which then uses a generator to convert this to electrical energy. The invention of new machines, from automated looms to steam engines, was the driving force behind the Industrial Revolution. In its turn, the drive to make better machines resulted in the development of the physics of thermodynamics. Initially conceived to improve steam engines, this provided the essentials to understand heat. The second law of thermodynamics tells us that all machines lose some energy as heat, so not all of it can be usefully employed.

DRILL DOWN | A "natural machine" may seem like an oxymoron, but all biological organisms contain components that play the role of machines. At the cellular level, complex molecules act as machines to carry out tasks, driven by chemical or electrical energy. Often these molecules are proteins, responsible for making the materials required for life—and breaking them up again. Other proteins are involved in machine roles, for example, in muscles, while the most complex biological machines, also featuring proteins, drive natural motors. Some bacteria have a filament extending from them called a flagellum. This is used to push the bacterium along, rotating like a propeller. The motion is driven by what is, in effect, a rotary motor.

WORK & ENERGY
Page 54

POWER
Page 56

LAWS OF THERMODYNAMICS
Page 76

FOCUS | *The accolade of the world's largest machine is given to the Large Hadron Collider (LHC) at CERN, the European particle physics laboratory based near Geneva. The LHC uses a 17-mile (27-kilometer) circular tunnel, running under the border of France and Switzerland. The collider incorporates 10,000 superconducting magnets and seven detectors, including the house-sized 14,000 tonne CMS detector.*

HEAT

THE MAIN CONCEPT | One of the most important developments of our understanding of energy was when the mysteries of heat were solved. For some time, the best theory of heat considered it to be a physical substance called caloric that flowed from hot to cold bodies. However, it became clear that heat was nothing more than the kinetic energy of the atoms and molecules in matter. The faster they moved, the more energy they had, the more heat there was. Usually, but not always, when heat is transferred from one body to another, it results in the recipient atoms and molecules speeding up. This understanding was essential in the development of better steam engines in the nineteenth century. The confusion over the nature of heat was not helped because it has three different ways to get from place to place. In conduction, there is direct contact between the atoms of one substance and the atoms of the other—it's transmission by collision. In convection, an intervening fluid—often air—carries the energy from place to place. And in radiant heat, the energy gets from place to place as infrared, an invisible form of electromagnetic radiation. The source atoms generate photons, which carry the energy even across empty space eventually adding to the energy of the atoms they later hit.

DRILL DOWN | One aspect of heat transfer that caused confusion was the way that transferring heat from one place to another did not always result in an increase in temperature (a measure of the energy of the molecules) in the recipient object. For example, when you heat water to 212°F (100°C), the temperature stops rising for a while, even though you continue to add heat. The same thing happens to a solid at its melting point. This is because at melting and boiling points, the heat is going into breaking bonds, instead of kinetic energy. Known as "latent heat," the bond breaking process interrupts the steady increase of temperature.

ATOMS
Page 20
KINETIC ENERGY
Page 58
HEAT ENGINES
Page 70

FOCUS | *Salford brewer James Joule performed a series of experiments to show that heat was equivalent to the energy used in doing mechanical work to produce it. His best-known experiment consisted of a weight dropping under gravity—so using a known amount of energy—which drove a paddle in an insulated container of water. As the paddle stirred the water, Joule measured the resultant rise in temperature.*

HEAT ENGINES

THE MAIN CONCEPT | In a world increasingly making use of electricity, the concept of a heat engine—which includes both steam engines and internal combustion engines—may seem increasingly outdated. Yet heat engines have played a huge role in the Industrial Revolution and the improvement of our lives. Before they were invented, we were limited to the work output of humans and animals, supplemented by waterwheels and windmills. Even if heat engines disappear from our streets, specialist forms of heat engines are likely to maintain a role amongst our portfolio of electricity-generation mechanisms. They take heat, which can be anything from geothermal heat to solar heating to heat from a nuclear reaction, and turn it into mechanical work. Although it's not necessarily obvious, what is happening within these heat engines is that heat is being transferred from a hot to a colder place via some kind of working substance (steam, for example, or the exhaust gases produced by burning a fuel) and along the way, that transfer involves doing work. The mechanism to transfer the kinetic energy of heat to work can vary considerably, although many use a piston or turbine blade to transfer kinetic energy from a gas to a rotating spindle—the drive shaft.

BONDING
Page 24
HEAT
Page 68
TEMPERATURE
Page 72

FOCUS | *The effectiveness of a heat engine depends on Carnot's theorem, devised by French scientist Sadi Carnot. This gives an absolute limit on the efficiency of a heat engine, based on the temperature difference between the heat source and the heat sink (the colder of the two). The bigger the temperature difference, the greater the efficiency of the engine, although it is impossible to achieve a perfect efficiency with no loss of energy.*

DRILL DOWN | The heat engines that aren't involved in turning heat into mechanical work use work to reverse the usual flow of heat. Instead of harnessing the energy of heat transferring from a hot place to a cold one, they take another source of energy and use it to transfer heat from a colder place to a hotter one. In general these are known as heat pumps, the best-known examples being refrigerators and air-conditioning units. Typically, heat pumps work by using energy (typically electricity) to compress a gas, condense it, then allow it very quickly to evaporate as it is pushed through a nozzle. The very rapid evaporation results in a sudden drop in the temperature.

TEMPERATURE

THE MAIN CONCEPT | Temperature gives us a feeling for how much heat there is in something, whether it's the atmosphere when we're talking about the weather, someone's body for medical reasons, or an oven when cooking. The temperature is a measure of the kinetic energy in the atoms and the molecules that make up something. It can come in the form of free movement, for example in a gas, or vibration. Individual atoms can also have different potential energy levels depending on the positions occupied by their electrons. Electrons around the atom can jump up to higher levels, storing energy until they drop back down. More pragmatically, temperature is what thermometers measure. The everyday scales we use are defined by set points of measurement, which tend to be arbitrary. So, the Fahrenheit scale, for example, was set with zero as the temperature of a specific mix of ice, water, and the chemical ammonium chloride. The size of the degrees was then fixed using the freezing point of water and the temperature of the human body. Celsius uses the more obvious freezing and boiling points of water as fixed points, with the range divided by 100 to give the size of a degree. However, science often uses the Kelvin scale. This is an "absolute" scale as it starts at the lowest possible temperature, absolute zero.

DRILL DOWN | German scientist Daniel Fahrenheit's scale, seems so strange, with its freezing point of water at 32 and boiling point at 212. The 0 point was selected as a mix of ice, water, and ammonium chloride was relatively easy to keep at a fixed temperature. The other two values used were the freezing point of water at 32 and human body temperature at 96. The reason for these values was that the difference between the temperatures was 64, and as this is a power of two, it's easy to draw the scale on a thermometer with 64 divisions by repeated halving.

ATOMS
Page 20

HEAT
Page 68

ABSOLUTE ZERO
Page 74

FOCUS | *The average temperature of the universe is often described as 2.73 K (that's −454.76°F or −270.42°C). This is a measure based on cosmic microwave background, photons that have been flying around the universe since it became transparent over 13 billion years ago. The value given is the temperature at which matter would give off this wavelength of light.*

ABSOLUTE ZERO

THE MAIN CONCEPT | For a considerable amount of time, the possibility of a limit on coldness was discussed. It was accepted that there should be one significantly before there was a logical explanation for such absolute zero temperature. In the eighteenth century, French natural philosopher Guillaume Amontons made an argument based on the reducing height of a column of mercury as temperature went down, but the detailed physical requirement for absolute zero to exist came from Scottish physicist Lord Kelvin, who based it on Carnot's work on heat engines. Once temperature is regarded as the energy in a material, then inevitably it must have a lowest value—an absolute limit, where there is no motion and all electrons are at their lowest possible energy levels. In practice, absolute zero cannot be reached, as it would breach the uncertainty principle of quantum physics, which says that energy levels of atoms will always have some fluctuation. Absolute zero also emerges from the concept of entropy, the amount of disorder in a substance, which is an important factor in heat and thermodynamics. Entropy measures the number of different ways the components of a system can be combined—the fewer the ways, the lower the entropy. At absolute zero everything would occupy a single state and so would have the lowest possible entropy.

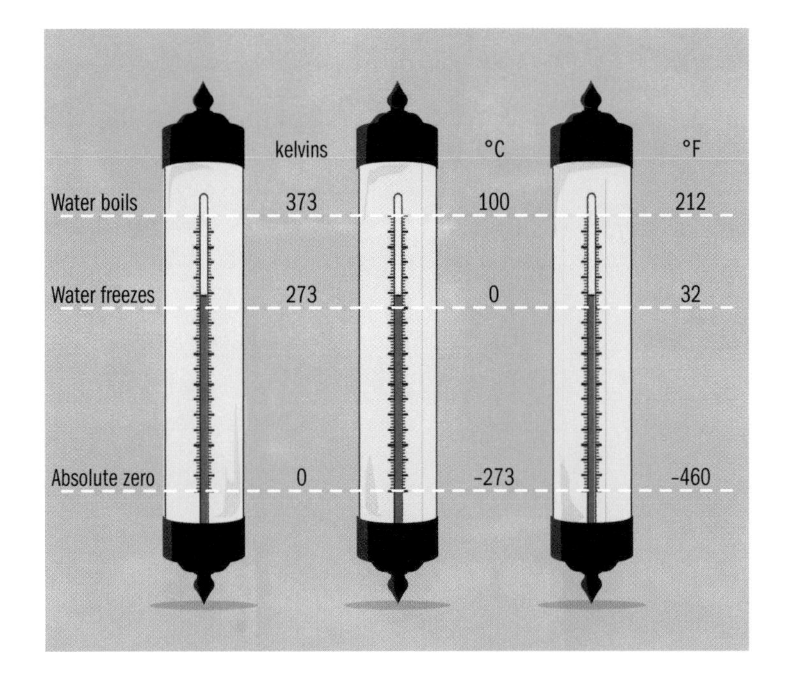

DRILL DOWN | Once absolute zero is known to exist, it seems inevitable that there should be a temperature scale based on it. The Kelvin scale, which uses the same sized units as degrees Celsius, known as kelvins (K), is the standard measure of temperature used in physics. It might seem that, by definition, since absolute zero is 0 K on this scale and it is not possible to be colder, then a negative Kelvin temperature should not exist. However, there is a strange state of matter that is said to have this. Such substances are not very cold but extremely hot, in a state where adding heat *reduces* the entropy in the system.

TEMPERATURE
Page 72
LAWS OF THERMODYNAMICS
Page 76
ENTROPY
Page 78

FOCUS | *The first person to get close to absolute zero was the Dutch physicist Heike Kamerlingh Onnes. In 1908, he became the first to liquefy helium, the element with the lowest boiling point, getting it down to 1.5 K (–456.97°F or –271.65°C), the lowest temperature achieved at that point. The latest attempts have reached temperatures of a billionth of a kelvin.*

LAWS OF THERMODYNAMICS

THE MAIN CONCEPT | Thermodynamics, the science of heat and its transfer, has a total of four laws. These run from the zeroth to the third, as there were originally three, and the zeroth law was a latecomer. The zeroth law states that two bodies are in thermal equilibrium if heat can flow between them, but doesn't. In reality, heat will flow back and forth, because different atoms will have more or less kinetic energy, but the net flow is zero. The first law deals with conservation of energy. It says that the energy in a system only changes to account for the work done between the system and the outside world and any heat that passes in or out of the system. An isolated system can neither gain nor lose energy. The second law is arguably most significant. It requires heat to pass from a hotter to a colder body. This can also be described as saying that the entropy (the disorder) in a closed system remains constant or increases. This happens as we move from having a hotter and a colder body to two bodies at the same temperature as before the process there was more order—hotter atoms in one body, colder atoms in the other. Finally, the third law tells us that it is impossible to reach absolute zero in a finite set of steps.

DRILL DOWN | The statistical nature of the second law is important. Imagine a simple experiment with a box divided into two, one half containing hot gas and the other cold. The second law tells us that if we remove the divider, the hot and cold gases will mix. The hot part becomes cooler and the cool part hotter. However, the atoms are not steered by anything. It is possible, if extremely unlikely, that having reached equilibrium, more of the hot, fast-moving atoms will happen to all end up back on one side. In such a case, heat has moved from a cooler to a hotter place and entropy has spontaneously decreased.

FOCUS | *You can see the zeroth law of thermodynamics in action when you take someone's temperature. The temperature on the thermometer is correct when it no longer continues to change. At that point, a thermal equilibrium is in place—there is no longer a net flow of heat between the patient and thermometer, so it is now showing the true value.*

HEAT
Page 68
TEMPERATURE
Page 72
ENTROPY
Page 78

ENTROPY

THE MAIN CONCEPT | Although entropy seems an abstract concept, it is central to thermodynamics. It is described as a measure of the disorder in a system. The higher the entropy, the more the disorder. The second law of thermodynamics tells us that entropy in a closed system will increase or stay the same. This is why it's easy to shatter a glass, but not to unshatter it. Similarly, it's easy to mix milk into coffee, but difficult to separate them. In each case, the natural inclination is to reach a more disordered state. It might not be so obvious why the system of separate coffee and milk is more ordered when they are mixed, but this is because when they are separate there are fewer ways for the molecules to be arranged—milk molecules can only be in one small volume, rather than anywhere throughout the cup. Disorder, and hence entropy, is measured by considering the number of different ways the components of the system can be arranged. Take the example of the letters on the page of a book. There are far more ways to arrange the letters randomly than to spell out the text of the book. So, a random arrangement has higher entropy than a specific piece of writing.

DRILL DOWN | Entropy and the way that it increases is sometimes used as an argument for a deity. On Earth, we see complex structures, such as organisms, which are far more ordered than a random collection of chemical elements. This, it is argued, shows divine intervention to break the second law. However, the second law of thermodynamics only applies to "closed systems." The Earth is open to a vast inflow of energy from the Sun—and using that energy makes it possible to reverse the second law and create order, at least temporarily. If we take the universe as a whole, though, we have a closed system. The expectation is that, over time, entropy will increase, leading to the universe eventually running down to an end state called Heat Death.

FOCUS | *The way that entropy naturally increases is cited as a reason for us thinking of time as something that has a natural direction of flow. Much of basic physics is reversible—it doesn't care which way time is considered to elapse. But the nature of entropy, which on average will always increase, is to suggest an "arrow of time" pointing in a particular direction.*

HEAT
Page 68

HEAT ENGINES
Page 70

LAWS OF THERMODYNAMICS
Page 76

"God runs electromagnetics on Monday, Wednesday, and Friday by the wave theory, and the devil runs it by quantum theory on Tuesday, Thursday and Saturday."

LAWRENCE BRAGG IN DANIEL J. KEVLES,
THE PHYSICISTS (1978)

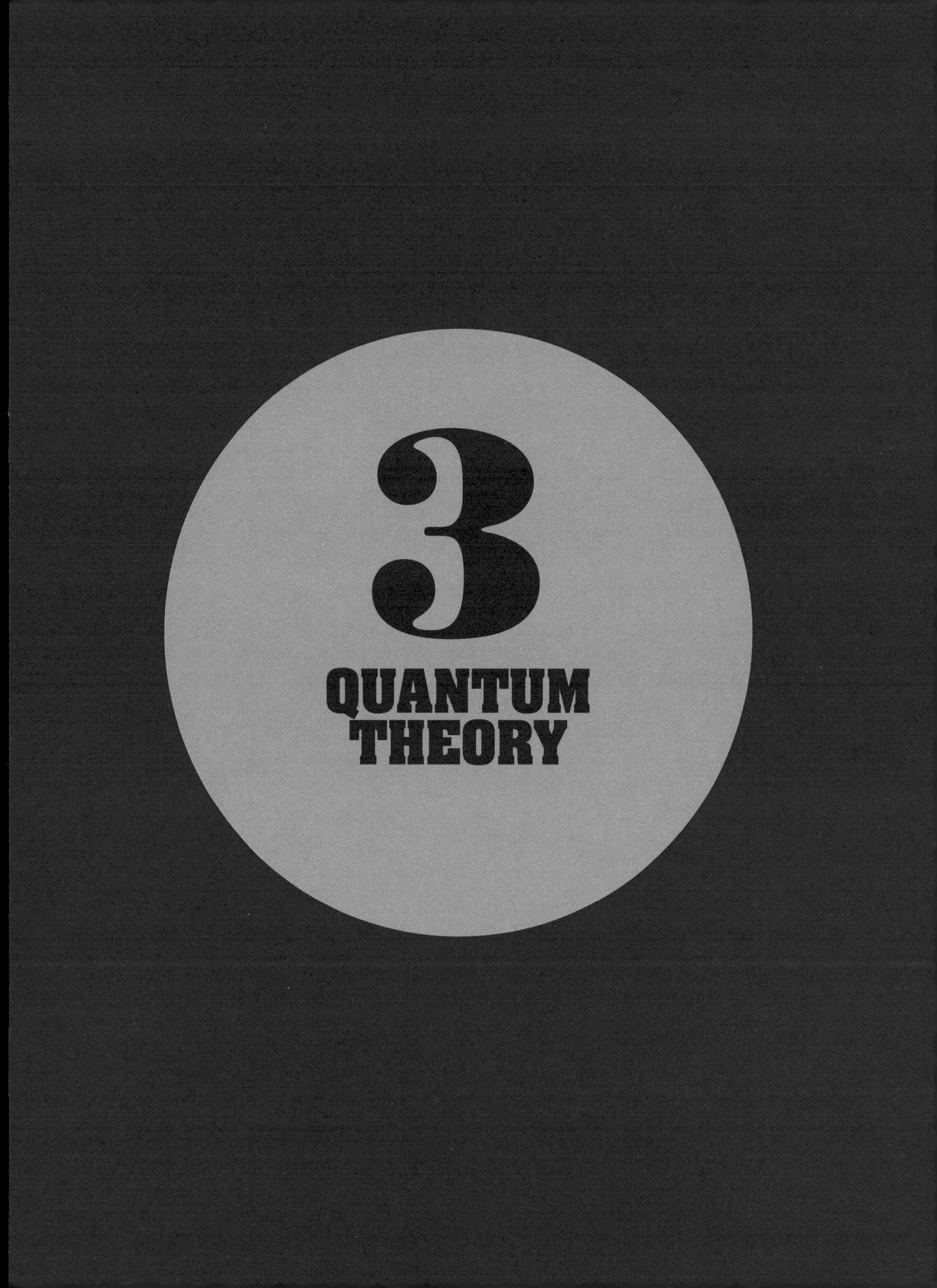

3 QUANTUM THEORY

INTRODUCTION

The basic concept of quantum physics sounds harmless enough. It's that most things in nature are quantized. Being quantized means coming in chunks, rather than continuous amounts. This is something we experience all the time. We know that matter is made up of separate atoms, rather than a continuous medium. A pack of salt or sugar is made up of separate grains. And, for that matter, money is quantized. It's impossible to spend 0.5124 of a cent. Yet when the quantum was brought into physics it caused a revolution.

This is because quantum theory was first applied to light—and in 1900, when this occurred, everyone was convinced that light was a continuous phenomenon that behaved as a wave. This wouldn't be the first time that scientists had to revamp their views of reality, of course. But there was a real problem that emerged from the idea that light was a collection of particles, quanta that would eventually be called photons. Waves are not localized; unlike particles, they spread out over space, and so can do things that we don't expect particles to do.

A quantum puzzle

An 1801 experiment, which had been taken as the definitive proof that light was a wave, demonstrated the problem superbly. Known as Young's slits, after the English polymath Thomas Young, the experiment passed light through two parallel narrow slits creating two distinct beams of light. The two beams of light interacted, combining to producing a wave phenomenon known as an interference pattern. Other waves do this, such as the ripples on a pond. If light were a stream of particles, one would expect to see the beams stay separate and not interfere. However, light is made up of particles, but they are particles that also act like waves. How can that be?

This puzzle and others like it persist to this day. Most notably, the apparent oddities of quantum physics arise from its use of probability. According to quantum theory, if a quantum particle doesn't interact with its environment, we can't say where it is. It doesn't even have a location, just a set of probabilities of where you might find it when you look. But these oddities have not resulted in the abandonment of quantum theory. If you don't worry about them and just work through the calculations, the theory is immensely effective at predicting what we will discover when interacting with the quantum world of the very small. Without quantum physics we would have no solid-state electronics, no lasers, no flat screen TVs, no LED light bulbs.

This led to a unique situation where quantum physics has ended up with a number of different explanations that try to bridge the gap between what is observed and what may be happening. The best accepted has been the "Copenhagen interpretation," which tells us to not worry about what is "really" happening, as we can never find out. In this approach there really is nothing except probabilities between observations. The approach is sometimes typified as "shut up and calculate."

Other interpretations try to get around the probabilistic nature. One suggests that a quantum particle always has a definite location, and that it can have wave-like properties because it has an associated "pilot wave" that guides it. Known as the Bohm interpretation after physicist David Bohm, this approach makes it necessary to do away with the concept of locality. Particles have to be able to communicate with others instantly at any distance. Another interpretation, the "many worlds" interpretation, incorporates the somewhat sci-fi concept that every time a particle could be in two different states, each exists in a different parallel universe.

Quantum behavior

The urge to produce these interpretations comes from an apparent mismatch between the quantum rules that govern tiny particles, such as atoms and photons of light, and the behavior of objects on the scale of which we are familiar, such as balls and people and cars. We know that a ball, for example, will follow a predictable path when thrown. Yet that ball is made up of quantum particles, each of which behaves in a very different fashion. It has been argued that we shouldn't think of quantum behavior as weird, because that's just how nature is. The only reason it appears to be weird is the collective behavior of quantum particles in those familiar objects seems quite different. However, it's hard not to be surprised—and fascinated—by the way that quantum objects behave.

Although quantum physics has been hugely successful both in predicting what is observed and in making possible the development of so much of our modern technology, there is a fly in the ointment. It is incompatible with that other great physics development of the twentieth century, Einstein's general theory of relativity. General relativity deals with the very large, as it explains the workings of gravity. But unlike the other forces of nature—electromagnetism and the strong and weak nuclear forces—the general theory does not allow gravity to be quantized. Even so, quantum theory is arguably our most successful physics theory to date.

BIOGRAPHIES

JAMES CLERK MAXWELL (1831–1879)

James Clerk Maxwell was the greatest physicist most people have never heard of. After studying at Edinburgh and Cambridge, he became a physics professor in Aberdeen, appointed at the age of 25, followed by professorships at King's College, London and Cambridge.

The two main planks of his work were in statistical mechanics and electromagnetism. He was one of the main contributors to the description of the behavior of gases through the statistical prediction of the interactions of molecules and put together the definitive equations describing the relationship of electricity and magnetism. His mathematics predicted that light should be an electromagnetic wave, a freestanding interaction between electricity and magnetism.

Maxwell's other great contribution was in moving from models of reality that were analogies with familiar things—his early work on electromagnetism modeled it using hexagonal wheels and small bearings—to purely mathematical models that had no direct relationship to familiar objects, an approach that proved essential for quantum theory. Maxwell's final position involved setting up the world-famous Cavendish Laboratory.

NIELS BOHR (1885–1962)

Born in Copenhagen in 1885, Niels Bohr's career was transformed by a year spent in the UK between 1911 and 1912. With a fresh doctorate, he set off to spend time in J. J. Thomson's lab in Cambridge but had a bumpy start when he criticized the Nobel Prize winner's book. Bohr had a more fruitful time following a transfer to Manchester to work under Ernest Rutherford.

It was in Manchester that Bohr assembled his ideas on a quantum structure for the hydrogen atom that he published in 1913, in so doing transforming ideas of how the atom worked. Bohr would be one of the central figures of quantum physics, establishing an Institute of Theoretical Physics in Copenhagen where many of the great names in the field would work.

At the institute, working with Werner Heisenberg, Bohr developed the "Copenhagen interpretation" of quantum physics and the concept of complementarity. The latter predicts that the way measurements are taken of a quantum phenomenon will influence the outcome—whether, for example, light acts like a wave or a particle. Bohr was not a great communicator, seeming to obscure as much as he explained—yet he would inspire a new generation of physicists.

ERWIN SCHRÖDINGER (1887–1961)

The Austrian physicist Erwin Schrödinger, born in Vienna in 1887, would have to serve out his military service in World War I before being able to take up an academic position. Working in Zurich in the 1920s, he developed an alternative approach to quantifying quantum properties to that of Heisenberg, portraying the behavior of quantum particles with a "wave equation," which would later be shown to provide the probability of finding a quantum particle in a particular location.

By 1927 Schrödinger had moved on to Berlin, but would soon leave Nazi Germany for England and then Ireland, where he would spend the remainder of his working life in Dublin. In 1935, in a short aside in a paper, he dreamed up his most famous creation, Schrödinger's Cat, a thought experiment designed to illustrate the oddity of quantum reality. While in Dublin he published an influential book, *What is Life?*, which inspired work on the structure of DNA.

Schrödinger had an unusual family life. Although he was married to his wife Anni for 40 years, all three of his children were by other women. He returned to Vienna in 1956 after his retirement, where he died in 1961.

RICHARD FEYNMAN (1918–1988)

American physicist Richard Feynman only had to speak a few words before his New York origins became clear. The Manhattan Project, leading to the development of the atomic bomb in World War II set his initial direction in physics. After working on the bomb at Los Alamos, New Mexico, he became a professor of theoretical physics at Cornell University, before moving in 1950 to his longest-serving academic home, Caltech in Los Angeles.

By the time he had moved to Caltech, Feynman had already done the work on quantum electrodynamics (QED) that would lead to him winning the Nobel Prize in physics in 1965. This theory established the quantum interactions between light and matter particles. However, it was as a communicator that Feynman achieved his greatest fame. His physics lectures, published in book form, became classics, while his collections of anecdotes, such as *Surely You're Joking Mr Feynman!*, won him many followers.

Feynman cemented his celebrity in his appearance on TV as a member of the panel investigating the *Challenger* shuttle disaster, where his use of ice water to demonstrate the flaw that caused the crash fascinated viewers.

1900

QUANTA

Max Planck suggests that light is given off as tiny packets of energy, known as quanta. Five years later, Albert Einstein assumes that Planck's quanta of light, later known as photons, are real to explain the photoelectric effect. This work would win both men a Nobel Prize.

BOHR'S MODEL

Niels Bohr produces a quantum model of the hydrogen atom.

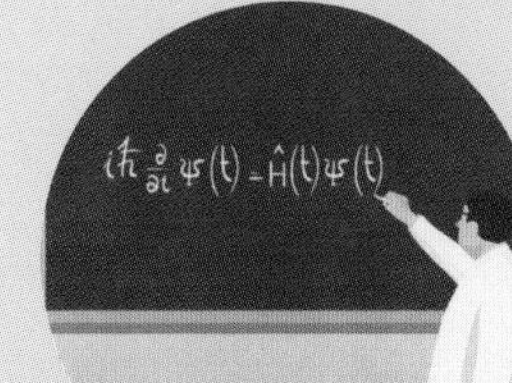

QUANTUM BEHAVIOR

Werner Heisenberg develops a more comprehensive mathematical description of the behavior of quantum particles called matrix mechanics. The following year, Erwin Schrödinger publishes an alternative approach to describing the behavior of quantum particles in the form of a wave equation.

1927

Heisenberg produces his uncertainty principle, which says that the more accurately we know one of a pair of linked properties of a quantum system, such as position and momentum, the less we can know about the other.

1928

ANTIMATTER

Paul Dirac produces an equation describing the behavior of electrons at relativistic speeds. To make his equation work, Dirac inadvertently devises the concept of antimatter, discovered four years later in cosmic rays by Carl Anderson.

1935

ENTANGLEMENT

Albert Einstein, assisted by Boris Podolsky and Nathan Rosen, describes the strange quantum behavior known as entanglement. Their paper attempted to show that quantum physics was incomplete because its predictions were so unlikely—but later experimental evidence proved Einstein and colleagues wrong.

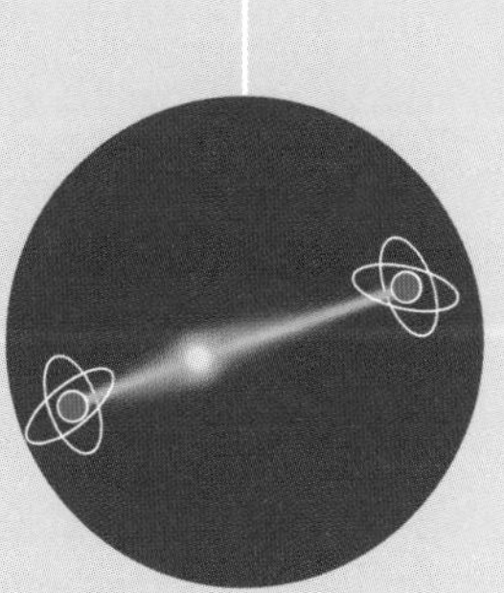

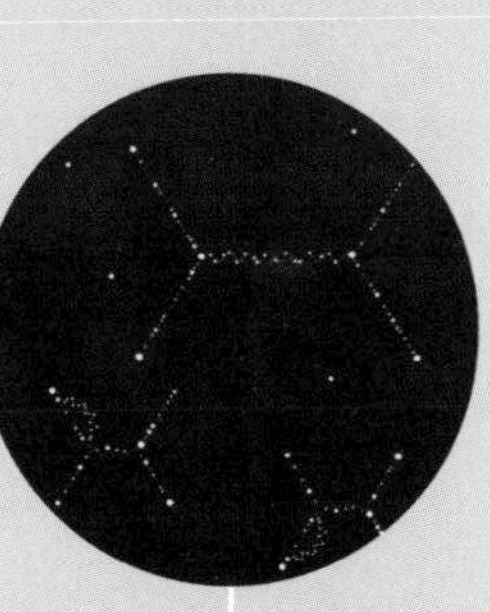

1948

QED

Richard Feynman, Julian Schwinger, and Shin'ichirō Tomonaga independently build on Dirac's work to develop quantum electrodynamics (QED), describing how all electromagnetic quantum phenomena take place. This theory explains the relationship of light and matter.

1973

QCD

Quantum chromodynamics (QCD), an equivalent of QED for the strong force between quarks is developed. This explains how quarks join to produce particles such as neutrons, protons, and mesons, and how the atomic nucleus holds together.

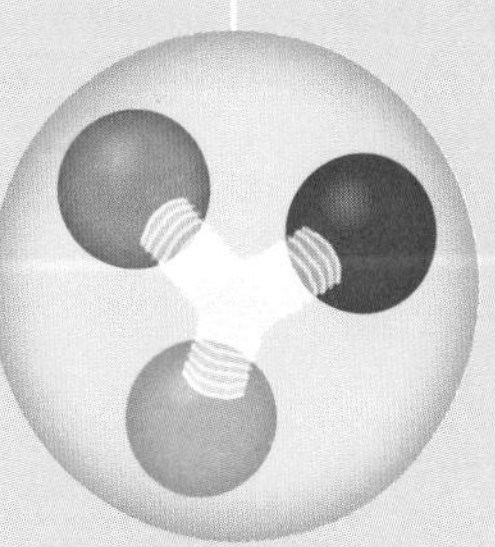

ELECTROMAGNETISM

THE MAIN CONCEPT | One particular form of energy, electromagnetic energy, forms a bridge between the topics of energy and quantum physics. In the 1860s, the Scottish physicist James Clerk Maxwell, unified the concepts of electricity and magnetism, which until then had been considered entirely separate phenomena. Others, notably Michael Faraday, had been aware of a connection between the two, but Maxwell produced a series of equations, eventually whittled down to four simple-looking mathematical statements that described how electricity and magnetism interacted. In developing his model of electromagnetism, Maxwell realized that it should be possible to produce a self-sustaining wave, where fluctuations in an electric field produced corresponding changes in a magnetic field which produced further oscillations in the electric field and so on. Such a wave could only exist at a single speed in any particular material, and when he calculated this speed, Maxwell discovered that it was the speed of light. After millennia of speculation, Maxwell had discovered what light was. The requirement for light to travel at this definitive speed was the inspiration that led to Einstein's special theory of relativity, and the secondary paper he produced on this, which showed that energy and matter were related in $E=mc^2$ (where c is the speed of light). Electromagnetic energy is the usual form that is produced when quantum particles interact and convert matter into energy.

SPEED OF LIGHT
Page 44
NUCLEAR ENERGY
Page 64
QUANTA
Page 90

DRILL DOWN | The separate concepts of electricity and magnetism were familiar long before it was realized what they were. Electricity was first studied in the static form, produced, for example, by rubbing amber to make it attract feathers and fluff, and perhaps make tiny sparks. The Greek name for amber is *elektron* and this is the root of the word "electricity." Meanwhile natural magnets known as lodestones had been discovered to point approximately to the North Pole. In the nineteenth century, current electricity became a more familiar concept, especially with the development of the electrical cell by Alessandro Volta. Before Maxwell's work, a number of discoveries had been made on how moving magnets generate electricity, or electrical currents in wires act as magnets, but it was Maxwell who pulled the whole together.

FOCUS | *James Clerk Maxwell was one of the first individuals who could legitimately be called a scientist rather than a natural philosopher. The word "scientist" was proposed by analogy with "artist" at a British Association for the Advancement of Science meeting in 1834, though it took some time to catch on. Other options included sciencist, sciencer, scientician, and scientman.*

QUANTA

THE MAIN CONCEPT | Maxwell's equations provided an essential foundation for quantum physics to be brought into being, but the starting point of this transformation was in 1900, when German physicist Max Planck introduced the concept of quanta. He did this in response to the "ultraviolet catastrophe." This refers to the way matter gives off light. The hotter an object, the higher the frequency the light has. We are familiar with this when, for example, a piece of metal is heated up. At room temperature it gives off invisible infrared light. As it gets hotter it produces a visible glow, first red, then yellow, and finally blue-white. This had been predicted by theory, except that the theory made it clear that at higher temperatures more and more blue, violet, and ultraviolet light should be given off. Even at room temperature bodies should glow dramatically in the ultraviolet, yet clearly this didn't happen. Max Planck tamed the catastrophe, producing exactly the distribution of light frequencies that was observed. But his solution came at a price. Light had to come in chunks, in quanta. Planck considered quanta a useful mathematical trick, but five years later Einstein was able to explain the photoelectric effect by assuming quanta were indeed real.

DRILL DOWN | Planck was never happy about his use of quanta, describing it as an "act of desperation." There had been good evidence since 1800 that light was a wave, an idea that was reinforced by Maxwell's work. Yet by saying that light was made up of quanta, Planck was regarding it as a stream of particles, which seemed a totally incompatible concept. Einstein, by contrast, was prepared to take the concept at face value. He realized that the photoelectric effect, where an electrical current is produced when some materials have light shone on them, behaves in a way that only makes sense if light were indeed a stream of particles.

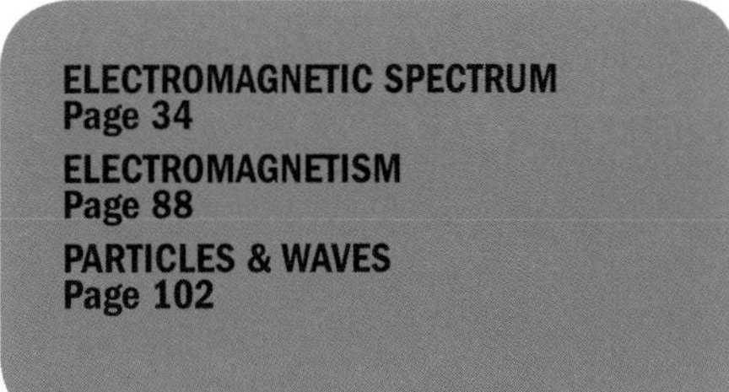

FOCUS | *The word "quantum," plural "quanta," just means an amount of something. This is the origin of terms such as "quantum physics" and "quantum theory." Something that is quantized cannot be obtained in continuous amounts, but comes in specific sized chunks. For example, cash is quantized. You can't have 3.614 cent or pence as coins, only the units that coins are available in.*

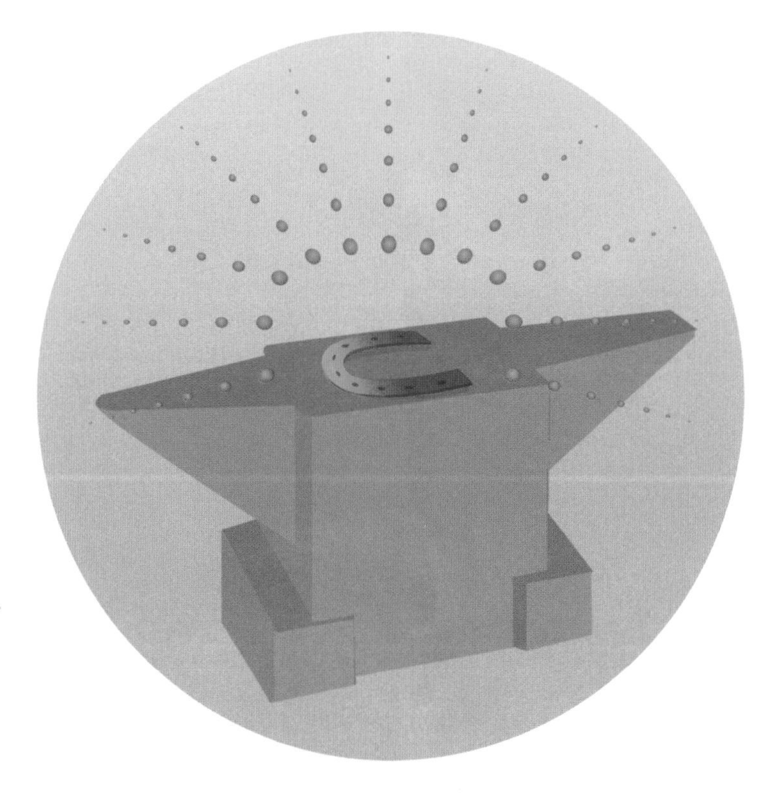

THE QUANTUM ATOM

THE MAIN CONCEPT | Once the electron was discovered as a component of the atom there had to be some kind of underlying atomic structure. J. J. Thomson, the electron's discoverer, suggested that electrons were scattered through a positively charged medium, but Ernest Rutherford and his team found that the atom had a small, positively charged core, which he called the nucleus. This meant that electrons had to be positioned outside the positive charge. A natural model for how this worked was the solar system, with electrons taking the place of the orbiting planets. Unfortunately, though this made a good picture, it couldn't work. When electrons are accelerated, they give off energy in the form of light. And bodies in orbit are constantly accelerating. All kinds of alternatives were tried, such as having the electrons static, balancing each other out, but nothing could be made to work. Then the young Danish physicist, Niels Bohr, inspired while working with Rutherford, came up with a quantum version of the atom. In this atom, electrons traveled around the atom as if they were on tracks. They couldn't drift into the nucleus. They could only move between these "tracks" in jumps known as quantum leaps. Each leap involved giving off or absorbing a photon of light.

DRILL DOWN | Although Bohr wasn't aware of it until well into developing his theory, the idea of his fixed possible orbits, with photons of light given off or absorbed as electrons jumped between them, explained an observation made by Swiss school teacher Johann Balmer in the year that Bohr was born. When a substance is heated it doesn't give off a continuous spectrum of light, but narrow bands of colors. Balmer had noticed that the frequencies of these colors for hydrogen fitted to a simple formula. Bohr realized this reflected the different energy levels that electrons could have according to his model (the tracks of Bohr's initial imagery), with the colors reflecting the energy of the photons given off.

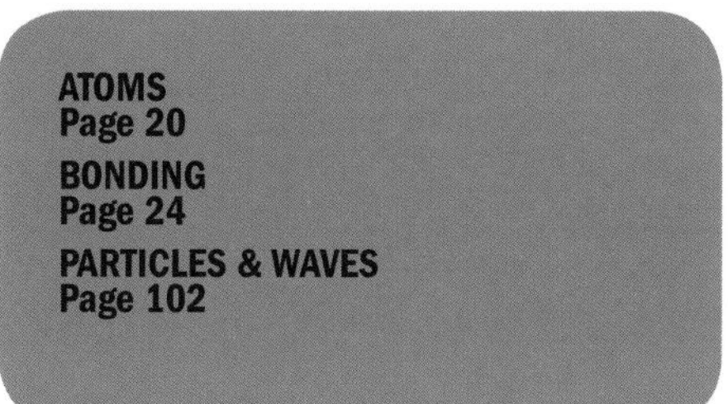

FOCUS | *Bohr might never have succeeded if he hadn't fallen out with an earlier professor. After winning his doctorate, Bohr had hoped to spend a year working at Cambridge with Thomson, but the pair failed to hit it off. A few months later, Bohr transferred to Manchester where Rutherford proved much easier to work with. It was in Manchester where Bohr began to formulate his atomic model.*

SCHRÖDINGER'S EQUATION

THE MAIN CONCEPT | The first developments in quantum physics were limited in their application—Bohr's model of the atom, for example, only worked properly for hydrogen. However, by the 1920s enough work had been done to produce a more general formulation of the behavior of quantum particles. The first attempt at this in 1925 was by German physicist Werner Heisenberg. Though his system termed "matrix mechanics" worked, it relied on using mathematical methods that physicists weren't familiar with, and it was totally abstract, simply a system of numbers for making predictions. The following year, Austrian physicist Erwin Schrödinger took a totally different approach to the same problem, using the more familiar mathematics of waves. Initially Schrödinger's wave equation caused a degree of confusion as it seemed to show that quantum particles would spread out to occupy more and more space over time. However, it was realized that instead, the square of the equation showed the probability of finding a particle in a particular location. This made more sense, even if it gave a challenging picture of reality, namely that quantum particles did not seem to have fixed locations until their position was measured. This concept made Schrödinger himself uncomfortable. Heisenberg's matrix mechanics and Schrödinger's wave mechanics were later shown to be mathematically equivalent and are collectively known as quantum mechanics.

DRILL DOWN | A significant aspect of Schrödinger's equation is the need to take the square of the value. The equation contains the mathematical symbol *i*, which represents the square root of minus one. Square numbers are arrived at by multiplying two identical numbers together, and that means they must always be positive. It is impossible to square a "real" number to give −1, and so mathematicians invented the unreal, or imaginary, number *i* to do it instead. A value which multiplied by itself equals −1 has proved very valuable in physics. By combining an imaginary value with a real value, e.g. *3 + 4i*, a "complex number" is constructed. Complex numbers can be used to represent locations on a plane and are useful for describing the behavior of something that changes with time, such as a wave. As the equation was squared, all the *i* values became real numbers again, so it did not produce an imaginary result.

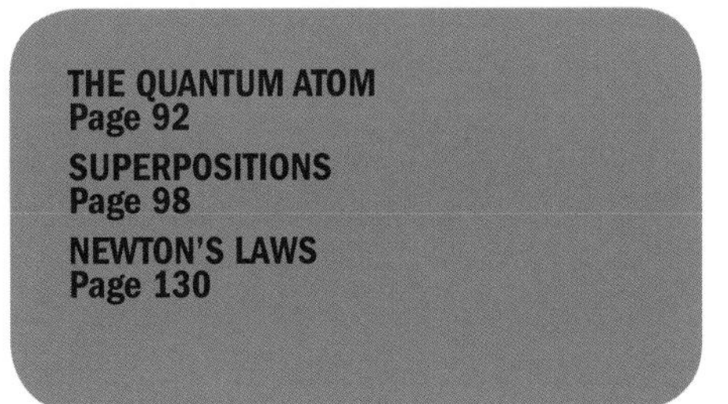
THE QUANTUM ATOM
Page 92
SUPERPOSITIONS
Page 98
NEWTON'S LAWS
Page 130

FOCUS | *Einstein's concerns about quantum physics were expressed in letters to Max Born, who came up with the idea that Schrödinger's equation described probabilities. Commenting on the concept that a quantum particle only had a probability for its location before observation, he wrote: "In that case, I would rather be a cobbler, or even an employee in a gaming house, than a physicist."*

THE UNCERTAINTY PRINCIPLE

THE MAIN CONCEPT | With the possible exception of Schrödinger's Cat, there is no other aspect of quantum physics so widely known (or so widely misinterpreted) as Heisenberg's uncertainty principle. It is sometimes represented as suggesting that nothing is definite, but in reality, it simply describes a relationship between pairs of properties of quantum particles or systems. The best known of these is between the position and the momentum of a particle. The more accurately we know one of these values at a particular point in time, the less accurately we can know the other. It's possible to measure either, but not both simultaneously. When Heisenberg first came up with the principle he mistakenly thought it was because the act of observing a particle—for instance, shining light on it—will change it. But it is an inherent part of quantum nature. A second pairing that is even more significant is energy and time. If we pin down a quantum system to a very narrow timescale, its energy can have a wide range of values. This means that even in empty space, on exceedingly narrow timescales, the energy present can reach levels high enough for "virtual particles"—pairs of matter and antimatter particles—to pop into existence for a minute fraction of a second then disappear again.

DRILL DOWN | Virtual pairs of particles that appear in empty space due to the uncertainty principle don't stay around long enough to be observed, so it may seem that this demonstration of uncertainty in action cannot be tested. However, this quantum-scale uncertainty can be seen in action on the large scale through the Casimir effect. This happens when two flat plates are very close to each other in a vacuum. Although there are no particles between or outside the plates, a tiny force is generated by virtual particles, which pushes the plates together. Usually, virtual particles do not push against their surroundings because they are equally likely to appear on one side as the other. But when the plates are very close together, fewer pairs form between the plates than outside them. As a result a tiny amount of pressure appears outside the plates, pushing them together.

TUNNELING
Page 100
THE VOID
Page 122
MOMENTUM & INERTIA
Page 126

FOCUS | *Physicists working at the quantum level take the uncertainty principle into account. Nowhere is this more obvious than in the world of particle physics. Early particle accelerators were quite small, but the Large Hadron Collider is a 17-mile (27-kilometer) ring. This is because to make precise enough measurements of position, the particles have to be given huge amounts of momentum, requiring a massive accelerator.*

SUPERPOSITIONS

THE MAIN CONCEPT | Absolutely central to quantum theory is the idea of superposition. This is often presented as meaning that a particle can be in more than one place at a time, or can have more than one value for a property such as spin. So, for example, in the Young's slit experiment that was used in 1801 to demonstrate that light was a wave, light is sent through two parallel slits, after which the beams of light interfere to cause a pattern. This phenomenon of interference also occurs with quantum particles, even if they are sent through the apparatus one at a time. So, it has been said that the particles "go through both slits." However, the correct view is that when not interacting with its environment, a quantum particle does not have properties such as position or spin. All that exist are probabilities. In the case of the Young's slit experiment, these probabilities are described by wave equations, and it is the probability waves that interact. When a quantum particle does not have specific values for properties, but rather a collection of probabilities, it is said to be in a superposition of the possible states it could be in. When it is observed, however, this superposition is said to collapse to a single observed value.

DRILL DOWN | The idea that a quantum particle is in a superposition of states was the inspiration for Erwin Schrödinger's thought experiment known as Schrödinger's cat. It employs a radioactive particle that will decay. We can't say when, just the probability of it happening in a particular timescale. The particle is therefore in a superposition of decayed and non-decayed states. Schrödinger envisaged an experiment where a detector released poison if the particle decayed, killing a cat. That means the cat was also in a superposition of dead and alive states, which seemed nonsense. In practice, the particle's interaction with the surroundings that are needed to trigger the detector would produce a single state.

TUNNELING
Page 100
PARTICLES & WAVES
Page 102
ENTANGLEMENT
Page 112

FOCUS | *When a particle is observed we never see it in both states at once. This transition from superposition to normality, known as collapse, is the most controversial aspect of the mainstream interpretation of quantum physics, as there is no clear mechanism for it. It is now often described as decoherence, where a particle interacts with its environment and no actual collapse is required.*

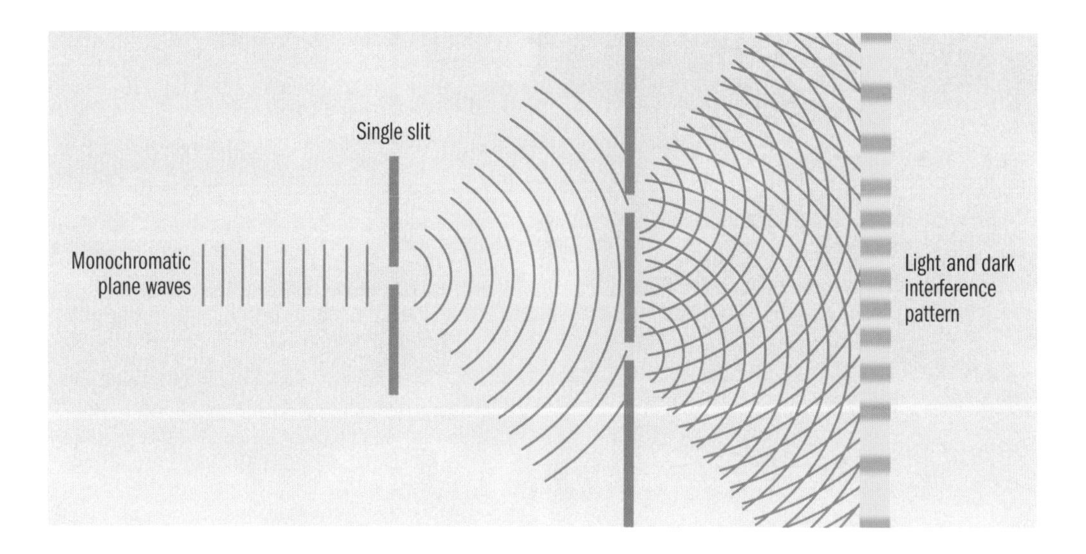

TUNNELING

THE MAIN CONCEPT | Although we use the familiar word "tunneling" in the phenomenon of quantum tunneling, it is a totally different concept to tunneling through an obstacle. When road builders construct a tunnel, they bore through an obstacle, providing a way to pass through. But in quantum tunneling, a quantum particle reaches the other side of an impassable barrier without traveling through it. We can see how this happens by thinking of the particle's position described by Schrödinger's equation. Over time, the position where we may find the particle spreads out according to the equation. If there is an obstacle in its way, there is a chance that it is already the other side of the obstacle. What is observed is that the particle was one side of the barrier and is now the other side of the barrier—it appears to have tunneled through but has not passed through the barrier itself, nor does it take any time to do the tunneling. This means that, for example, when a photon of light tunnels through a barrier, it appears to be traveling faster than light. Tunneling is a widely observed phenomenon, whether it is taking place in the depth of the Sun to enable fusion to take place or in electronic devices found in most homes.

SPEED OF LIGHT
Page 44
SCHRÖDINGER'S EQUATION
Page 94
SUPERPOSITIONS
Page 98

DRILL DOWN | The Sun is the biggest example of the importance of tunneling in our neighborhood. For a star like the Sun to function it has to be able to squeeze hydrogen nuclei (protons) sufficiently close together to fuse to form helium nuclei. However, even the temperature and pressure of the Sun's core isn't enough to overcome the repulsion of the positively charged protons. It is only because they can tunnel through the barrier of the repulsion that the Sun works. Back home, tunneling is used in the flash memory of thumb drives, phones, and laptop computers. This memory keeps data when the power's off by holding it in an insulated store, only accessible by quantum tunneling.

FOCUS | *Superluminal experiments send photons faster than the speed of light by allowing them to tunnel through a barrier. Because the photon spends no time in the barrier, its end-to-end time is faster than light speed. Some physicists claimed no information could be sent this way, but Austrian scientist Günter Nimtz sent a Mozart symphony at over four times the speed of light.*

PARTICLES & WAVES

THE MAIN CONCEPT | At the heart of the deviation of quantum physics from classical physics—and, for that matter, our experience of the world around us—is the way that a quantum object can act as a particle or a wave, depending on what we do with it. At the start of the twentieth century some things were considered waves—light, for example—while others, such as electrons and protons and atoms, were particles. Max Planck and Albert Einstein chipped away at this distinction, showing that light could behave as a stream of particles. French physicist Louis de Broglie turned the concept on its head in 1923, suggesting that particles could equally behave like waves. Within a few years, wave-like behavior, such as diffraction and interference, was demonstrated in electrons. Not only would Niels Bohr's quantum atom model be reinforced by this possibility, by the late 1920s, Bohr and Werner Heisenberg had formulated their "Copenhagen interpretation" of quantum physics, including the principle of complementarity. This said that a quantum entity could act as a wave or as a particle, but not both simultaneously. For example, if a detector that pins down the paths of individual particles is put in place when electrons are set up to cause an interference pattern, the pattern vanishes.

DRILL DOWN | When Bohr devised a quantum model of the atom, he suggested that electrons could only occupy fixed track-like orbits. With the idea that particles such as an electron could be wave-like, there was justification for the size of the orbits, which would need to be the right size for an electron's wave to fit exactly around the orbit, so the wave matched up when it returned to its start. The basic orbits electrons occupy are known as shells, but the statistical distribution of an electron around the atom, determined by Schrödinger's equation, is called an orbital. Each shell can have subshells, each of which can have more than one possible orbital.

ELECTROMAGNETIC SPECTRUM
Page 34
THE QUANTUM ATOM
Page 92
SCHRÖDINGER'S EQUATION
Page 94

FOCUS | *The concept of quantum particles or waves is a model. In physicist's term, a model is an analogy of how reality behaves. The analogy may have conceptual elements but will almost always have mathematics at its heart. When we say light is a wave, or a stream of particles, or a disturbance in a quantum field, these are useful models, but light is nevertheless just light.*

NUCLEAR FORCES

THE MAIN CONCEPT | Nature has four fundamental forces. Two can be easily seen in action—gravity and the electromagnetic force, which holds matter together and stops you passing through your chair when you sit on it. The other two (the strong and weak nuclear forces) are obvious at the quantum level. Of these, the strong nuclear force is the most dramatic. Its primary role is holding together quarks, the fundamental particles that make up protons and neutrons in the atomic nucleus. The strong force is carried by particles appropriately called gluons, the equivalent of the photons that carry the electromagnetic force. The strong force is a unique force because, over a very close range, it gets stronger as you separate quarks, which means they are never seen isolated. Enough of the strong force leaks from protons and neutrons to be able to hold together the nuclei of atoms, even though the protons are repelling each other because they all have a positive electric charge (and like charges repel). The weak nuclear force typically changes one type of a particle into another type, for example changing a "down" quark into an "up" quark, which results in a neutron turning into a proton, giving off a pair of other particles in one form of nuclear decay. Three different particles can carry the weak force: W^+, W^-, and Z bosons.

ATOMS
Page 20
MASS
Page 22
THE STANDARD MODEL
Page 106

DRILL DOWN | When a particle is described as "carrying" a force, a stream of the carrier particles is exchanged between the two particles attracting or repelling each other. In the case of the strong force, its gluon carriers have a "color charge," which unlike the electrical charge has eight different values, which determine what type of quark it will act on. By contrast, the carrier particles of the weak force have more familiar electrical charges: the W^+ is positive, the W^- negative, and the Z is neutral. The other oddity of the weak force carrier particles is that unlike photons and gluons, they have mass—they're around 100 times as massive as a proton.

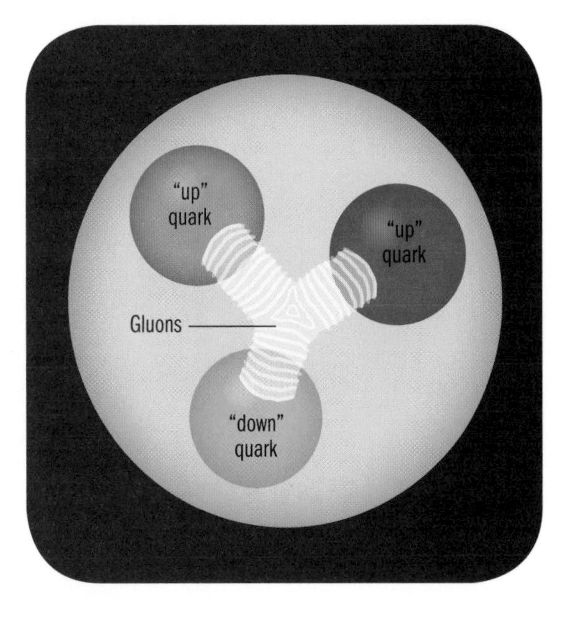

FOCUS | *As the names suggest, the weak force is weaker than electromagnetism, while the strong force is stronger than electromagnetism. If, for example, the electromagnetic force were stronger than the strong force, atomic nuclei would not stay together. There is so much energy in the strong force that it contributes the bulk of the mass of protons and neutrons—and therefore almost all matter—via the relationship* $E=mc^2$.

THE STANDARD MODEL

THE MAIN CONCEPT | During the 1960s it seemed as if what had been a simple set of fundamental particles was becoming a messy, complex "particle zoo." New particles seemed to be being discovered every few weeks. However, by the end of the 1970s, a "standard model" describing the basis of matter and forces with a relatively small set of fundamental quantum particles was agreed. This originally featured four quarks: up, down, charm, and strange, though top and bottom were later added; six particles known as leptons: the electron, muon, and tau, each of which had an equivalent neutrino; four "gauge bosons," which are carriers of fundamental forces: photon, gluon, Z and W bosons; and the Higgs boson, which was responsible for some particles unexpectedly (according to theory) having mass. Each matter particle also has an antimatter equivalent, as does the charged W boson, while the uncharged bosons are often described as being their own antiparticle. This relatively small set of particles makes up all the known matter, with the exception of dark matter (if it exists). In addition it explains all interactions between matter particles, except for gravity. The best current theory of gravity, the general theory of relativity, is not a quantum theory so does not contribute to the standard model. If a quantum theory of gravity were found, the carrier boson would be called a graviton.

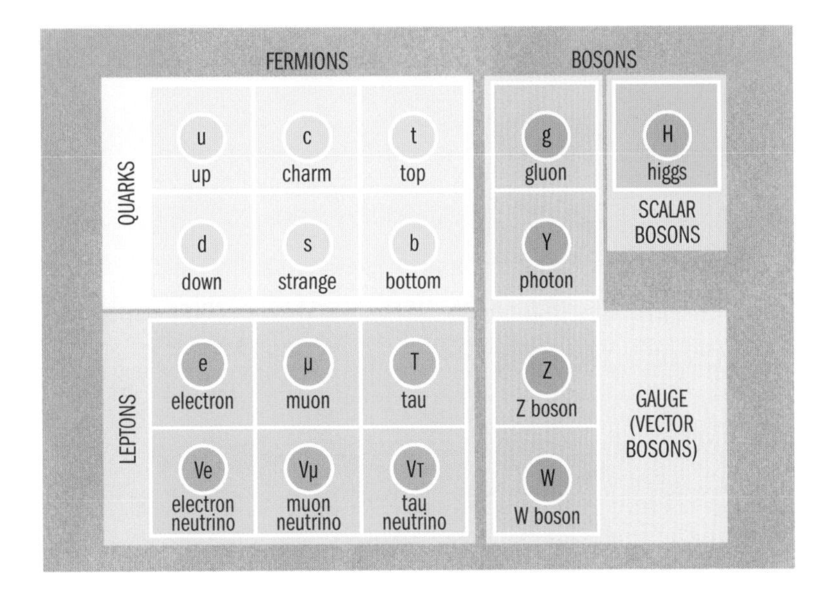

DRILL DOWN | The particles in the standard model divide into two distinct types, fermions and bosons. All matter particles are fermions, as are the insubstantial neutrinos. All force carrier particles are bosons. The names refer to the physicists Enrico Fermi and Satyendra Bose. Fermions obey "Fermi-Dirac statistics," while bosons obey "Bose-Einstein statistics." In practice this means that fermions are restricted by a quantum law of nature known as the Pauli exclusion principle, which only allows two fermions (such as the electrons in an atom) to be in the same quantum system if they have different values for certain properties such as spin. Bosons, by contrast, don't obey the exclusion principle and happily crowd together.

ATOMS
Page 20

QUANTA
Page 90

FIELDS
Page 108

FOCUS | *The fundamental matter particles called "quarks" were named by American physicist Murray Gell-Mann, beating an alternative proposal to call them "aces." Gell-Mann intended the name to be pronounced "kwork," but spelled it with an "a" after seeing James Joyce's line "Three quarks for Muster Mark" in* Finnegans Wake. *This seemed appropriate, since quarks combine in threes to form protons and neutrons.*

FIELDS

THE MAIN CONCEPT | Ever since English scientist Michael Faraday came up with the concept in the mid-nineteenth century, the idea of a field has been important in physics. It would take on extra significance in the twentieth century, as fields proved essential in understanding quantum interactions. A field is something with a value throughout space (and usually time). Those values can differ from location to location and from time to time. A contour map plots a kind of field, showing the height at each point. Fields proved immensely valuable in understanding how a force of nature worked. Faraday was no mathematician and used fields qualitatively. He had seen the way that iron filings produce a contour map from pole to pole of a magnet, denoting the "contours" of the magnetic field, which he named "lines of force." The closer together those lines were, the stronger the field. Scottish physicist James Clerk Maxwell made field theory mathematical, explaining how electricity and magnetism interacted. He showed that light was effectively a ripple in the magnetic field, causing a ripple in the electrical field and so on. Quantum field theory considers waves and particles as disturbances in quantum fields. An additional quantum field, the Higgs field, was predicted in the 1960s for being responsible for providing some particles with mass.

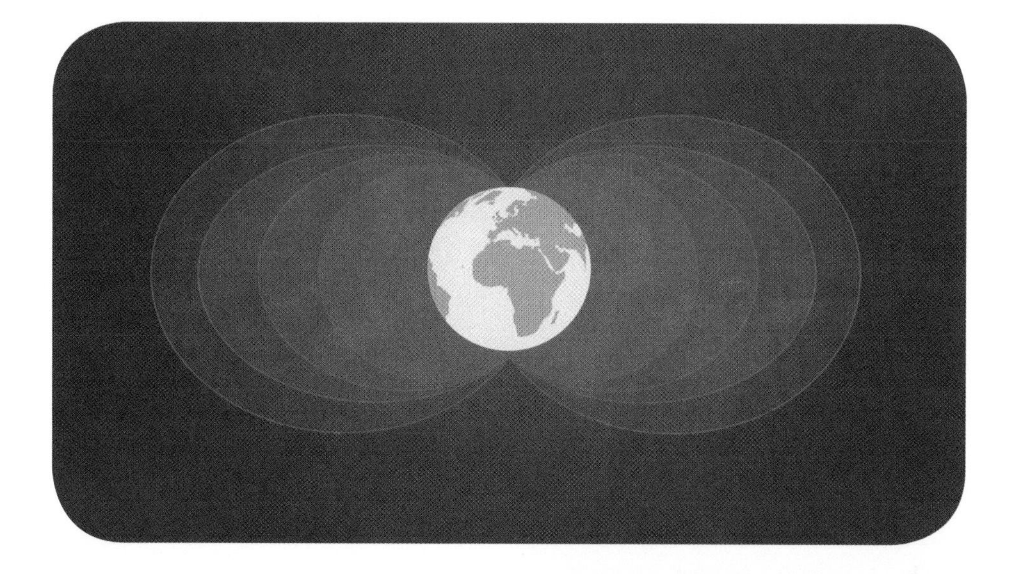

DRILL DOWN | Quantum theory predicted that force-carrying bosons should have no mass, as is the case with the photon and the gluon. Unfortunately, the bosons responsible for the weak nuclear force do have mass. A theory developed by six physicists, working in three groups—Robert Brout, François Englert, Gerald Guralnik, Richard Hagen, Peter Higgs, and Tom Kibble—suggested this mass was caused by an extra field, known as the Higgs field. If this field existed, it should produce its own boson, the Higgs boson, which was detected at CERN in 2012.

ELECTROMAGNETIC SPECTRUM
Page 34

ELECTROMAGNETISM
Page 88

THE STANDARD MODEL
Page 106

FOCUS | *The Higgs boson was nicknamed "the God particle," which made newspaper headlines, but irritated physicists. This name had nothing to do with the particle's importance. American physicist Leon Lederman introduced the phrase in 1993. He had wanted to call his book on the search for the particle "The Goddamn Particle" as it was difficult to find. His publishers wouldn't allow this and made it* The God Particle.

THE MAIN CONCEPT | Matrix mechanics and Schrödinger's equation provided the basics of quantum physics, but they did not deal with all the complexities of interactions between matter and light. Similarly, Maxwell's work on electromagnetism worked well, but predated Einstein's special relativity, which showed that if something was moving quickly—as quantum particles often do—Newton's laws of motion weren't good enough. In 1928, the British physicist Paul Dirac put together a description of the behavior of electrons that incorporated special relativity, making the first step into what would become known as quantum electrodynamics, or QED. Later work, undertaken independently by American physicists Richard Feynman and Julian Schwinger, and Japanese physicist Shin'ichirō Tomonaga filled out the detail of QED to make a complete theory of the interactions of light and matter. A significant part of Feynman's contribution was the Feynman diagram, which is both a way to illustrate interactions of quantum particles and used in quantum calculations. Initially there were concerns about QED as it depends on thinking of particles traveling every possible route and interacting every possible way, which can lead to calculations producing an infinite answer. A process known as renormalization was adopted, where actual values, for example of the electron's mass, were substituted for the infinite outcome. The result was a remarkably accurate theory.

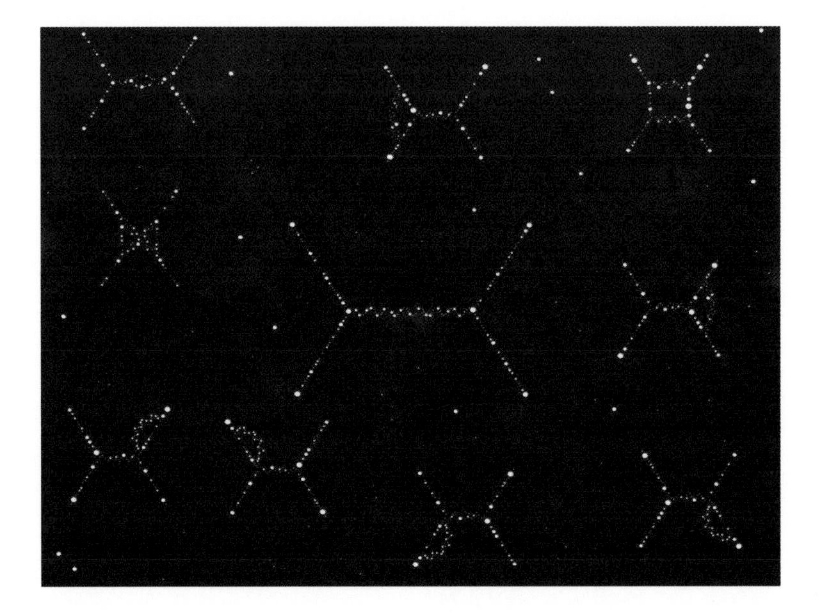

DRILL DOWN | Feynman diagrams are visual equivalents of the mathematics used in QED. They are plots where one axis is time and the other space. The movements of matter particles are shown by straight lines, while photons are seen as wiggly lines. Whenever lines connect, the particles are interacting. Typically, multiple diagrams will be drawn to cope with the alternative routes and methods of interaction that are possible in a quantum event. Since the diagrams were first used, they have also been deployed in QCD (quantum chromodynamics), the equivalent of QED for interactions of quarks and gluons. Richard Feynman had a van painted with the diagrams which he used to drive around the CalTech campus.

FOCUS | *Of all the theories of physics, QED is the most accurate in its predictions when compared with the results of experiments. As Richard Feynman pointed out, the calculation for the strength of electromagnetic interaction between particles is within one ten billionth of experimental results, equivalent to measuring the distance from New York to Los Angeles to the width of a hair.*

ANTIMATTER
Page 30
THE STANDARD MODEL
Page 106
SPECIAL RELATIVITY
Page 138

ENTANGLEMENT

THE MAIN CONCEPT | Although he was a founder of quantum physics, Albert Einstein grew increasingly uncomfortable about it, particularly the idea that when a particle wasn't observed it was nothing more than probabilities. In 1935, along with two younger physicists, Boris Podolsky and Nathan Rosen, he wrote a paper titled "Can Quantum-Mechanical Description of Physical Reality Be Considered Complete?" This detailed the consequences of a concept predicted by quantum theory known as entanglement. In the paper, Einstein described a thought experiment where two quantum particles are created in a linked state where they are described as entangled. If this were the case, according to quantum theory, you could separate the particles to opposite sides of the universe and a change in one particle would be instantly reflected in the other, apparently contrary to special relativity. As far as Einstein was concerned, this showed quantum physics was flawed. He proposed that the concept that quantum particles only had probabilities of properties was wrong, and instead they had preset (although unknown) values known as hidden variables. The alternative was that one particle could influence another instantly, which Einstein disparaged as "spooky action at a distance." Entanglement was finally demonstrated to have its "spooky" ability in the 1980s. Since then it has been used to produce unbreakable encryption and quantum teleportation.

DRILL DOWN | It might seem that the instant link of entanglement could carry messages, but the "information" it sends is random. Take an entangled pair of particles, each with a 50:50 chance of its spin being up or down. If one is examined and the spin is up, the other instantly has a spin of down—but there was no way to control this. However, this randomness is a benefit when using this mechanism to communicate an encryption key. Entanglement also makes it possible to transfer properties from one particle to another, a process known as quantum teleportation. This is essential in building quantum computers, which use quantum particles instead of conventional switch-based components for calculations.

SCHRÖDINGER'S EQUATION
Page 94
SUPERPOSITIONS
Page 98
SPECIAL RELATIVITY
Page 138

FOCUS | *Quantum computers are still at the experimental stage, but work is going on in labs around the world. When they can be made to work fully, there are already two significant algorithms to use on them. One would make it possible to crack most current internet encryption. The other allows a computational search to be carried in a fraction of the steps currently needed.*

“Today scientists describe the universe in terms of two basic partial theories—the general theory of relativity and quantum mechanics. They are the greatest intellectual achievements of the first half of this [twentieth] century.”

STEPHEN HAWKING,
A BRIEF HISTORY OF TIME (1988)

4

MOTION & RELATIVITY

INTRODUCTION

The mechanics of motion has been at the heart of physics ever since it was first described. However, it took Galileo to realize how important relativity was to understanding it, and Einstein to complete the picture. The ancient Greeks largely had a back to front understanding of movement, but one that makes perfect sense given their experience of the world. They saw that something moved when it was pushed and stopped moving when it was no longer pushed. They saw that a stone fell faster than a feather. And so, without testing this in any kind of structured way, they came up with a description of the physics of motion that was wrong, yet appealed to common sense.

Ironically, Aristotle pretty much hit on Newton's first law of motion around 2,000 years before Newton, but only as an argument to show why he thought that a complete void was impossible. If there were a vacuum, Aristotle said in his *Physics*, "No one could say why something moved will come to rest somewhere; why should it do so here rather than there? Hence it will either remain at rest or must move on to infinity unless something stronger hinders it." He thought this was a ludicrous consequence, yet it's pretty much exactly what Newton's first law tells us.

Galilean relativity

How Galileo tore apart the ancient Greek idea that heavier things fall faster, was not by dropping balls off the Leaning Tower of Pisa as legend has it (there is no evidence he did this), but with a neat thought experiment, worthy of Greek philosophy. This considered what would happen to two weights falling while connected by a rope to make them a single object. However, his most impressive insight was into relativity. The idea that movement is not something absolute, but is only significant when considered relative to something else (known in physics circles as the frame of reference).

How we describe movement is always relative. Even scientists sometimes forget this. Some while ago, my old Cambridge college posted a picture on Facebook showing the college chapel at night with the stars appearing as circular smears behind it. "This is Laurence Moscrop's shot of the chapel by night – with the stars moving behind it," commented the college. Someone soon commented "Earth rotating, I think, rather than the stars moving… we should be clear as to what it is illustrating!" The college *was* clear. It's certainly true that from a vantage point outside the Earth, best suited to studying the orbits of the planet, we would see the stars not moving and the Earth rotating. But from the viewpoint on the surface of the Earth, the one we conventionally use when talking about cars moving at 50 miles per hour, or people being stationary when sitting down, the Earth *isn't* moving. The stars are. Motion is relative. Galileo pointed out that if we were in an enclosed boat, traveling at a steady speed with no windows, it would be impossible with the technology of the day to determine with any physics experiment that the boat was moving. As far as the objects on it are concerned, it isn't moving. (Admittedly rotating things are more complex than Galileo's boat, but considering frames of reference is the essence of relativity.)

It's perhaps strange, then, that if you stopped someone in the street and asked them which name they associated with relativity they would almost universally say Einstein. This is not to say that Einstein's work wasn't remarkable—it was. But Galileo got there first. The whole business about the Earth traveling around the Sun (from the right frame of reference) that he is largely remembered for was a minor part of his output, based on other people's earlier arguments. It was in the physics of movement that Galileo did his greatest work.

Special relativity

What Einstein added came in two separate chunks. Firstly, he combined Galilean relativity and Newton's laws of motion with a revelation that had been highlighted by James Clerk Maxwell. In working on electricity and magnetism, Maxwell had demonstrated that light was an electromagnetic wave, because such a wave could *only* travel at the speed of light in any particular medium. Einstein realized that this meant that light could not be subject to Galilean relativity. Whichever way you move with respect to it, light always has the same velocity, or it would disappear. Plugging that fact into the laws of motion delivered the remarkable effects of special relativity.

Special relativity improves on the Galilean version for steadily moving objects. It produces almost identical results for most normal speeds, but get up to a reasonable fraction of the speed of light and strange effects kick in. Einstein then went on to the general theory of relativity, which added in the effects of gravity and acceleration (which Einstein realized were equivalent). Taken together, his work transformed our understanding of motion, relativity, gravity, and the workings of the universe.

BIOGRAPHIES

GALILEO GALILEI (1564–1642)

The son of a musician with scientific leanings, Galileo Galilei was born in Pisa, Italy in 1564. Galileo was due to follow his uncle into medicine, but after two years at university switched to mathematics. His early work while a professor at the universities of Pisa and Padua was on motion. However, Galileo was always looking for new opportunities. When in 1609 an early telescope was brought to Italy from the Netherlands, he hurried to construct his own device and, with the help of a confederate to delay his rival's instrument, got his telescope to Venice first.

As well as selling the device for nautical use, he made studies of the Moon and planets, discovering four of Jupiter's moons. Inspired by his observations, he began to support arguments from Copernicus on planetary motion, publishing his book *Dialogue Concerning the Two Chief Systems of the World*. Although he had obtained the Church's support for the book, the approach he took was considered too supportive of Sun-centered astronomy (and insulted the Pope). Galileo was tried and kept under house arrest for life. After his trial he wrote his masterpiece, *Two New Sciences*, summarizing his work on matter and motion. Galileo died in Florence in 1642.

ISAAC NEWTON (1643–1727)

Isaac Newton was born in 1643* at the family farm in Lincolnshire, England. He went to Cambridge in 1661, but soon after graduating, in 1665, the university closed as plague broke out. He spent 18 months at home, where he claimed he came up with many of the ideas that made him famous. After returning to Cambridge, he became the Lucasian Professor of Mathematics in 1669. Two years later he demonstrated a new reflecting telescope to the Royal Society and was elected a Fellow.

Newton followed up with a letter to the Society on light and color, which was criticized—a prickly Newton withdrew from the scientific mainstream until the 1680s, when astronomer Edmond Halley persuaded him to consider the motion of the planets. Halley published Newton's masterpiece on motion and gravity *Philosophiae Naturalis Principia Mathematica* (Mathematical Principles of Natural Philosophy). In 1696, Newton became Warden of the Royal Mint and engaged little further in science, apart from publishing *Opticks* in 1704 based on work from decades earlier. He died in London in 1727 aged 84.

** Modern dating. Newton's birthdate is often given as Christmas Day 1642, but that is old style dating, where his death year was 1726.*

ALBERT EINSTEIN (1879–1955)

Born in Ulm, Germany, in 1879, Einstein proved a rebellious youth, disliking the rigid school system and refusing to accept the requirement for German citizens to undertake national service. At the age of 16 he rejected his German citizenship and moved to Switzerland. On his second try, he got into the prestigious Zurich Polytechnic, but attended few lectures and only scraped a pass. Unable to get an academic job, he took the post of clerk in the Swiss Patent Office. While working there in 1905 he published four remarkable papers, establishing the reality of atoms, showing how the photoelectric effect required light to be quantized (for which he won the Nobel Prize), establishing special relativity and showing that $E=mc^2$.

By 1909, his work was beginning to be recognized and he got his first academic posting, reaching a professorship in 1914. The next year he published his masterpiece, the general theory of relativity which explained gravity, making him a worldwide star. As Nazi Germany became a hostile environment, he moved to the US in 1933, taking up a position at the newly formed Institute for Advanced Study in Princeton, New Jersey. He remained there until his death, aged 76, in 1955.

STEPHEN HAWKING (1942–2018)

The best-known physicist of the late twentieth and earliest twenty-first century, Stephen Hawking was born in Oxford, England, in 1942. He attended Oxford University before moving to Cambridge for his doctorate in 1962. He would remain at Cambridge for the rest of his working life, apart from a five-year stint at Caltech in the 1970s. Hawking specialized in the general theory of relativity – Einstein's theory on the nature of gravity – and phenomena arising from it, notably black holes. His most noted piece of work was the prediction of the existence of Hawking radiation, but as it is unlikely to ever be observed, his work was not in the running for the Nobel Prize.

Although he continued to work on theory, Hawking took on a second role as a science communicator with the unexpected success of his book, *A Brief History of Time*. Despite the book's reputation of being left unread by many purchasers, it proved influential both in the development of popular science and the careers of many physicists. The publicity surrounding Hawking's survival of motor neuron disease far longer than his expected lifespan enabled him to reach out beyond the scope of most scientists. He died in 2018, aged 76.

TIMELINE
FORMATION OF THE UNIVERSE

SECONDS LATER
Inflation stops, leaving the universe expanding more sedately. The early universe goes through rapid phases of particles annihilating to pure energy. 10 seconds in, the universe is primarily energy (photons of light), which then produce ionized matter.

STARS
The first stars begin to shine, around 100 million years after they first began to form from gas and dust, pulled together by gravitation until they had sufficient mass to enable nuclear fusion to begin.

13.6 billion years ago

BIG BANG
The space and time of our universe comes into existence. An immeasurably small fraction of a second later the big bang occurs and the universe begins to expand. 10^{-33} seconds in, the universe undergoes inflation, expanding to become 10^{26} times bigger.

ATOMS FORM
The universe becomes transparent as ions become atoms. The first light to pass through it is detectable to this day as the cosmic microwave background radiation.

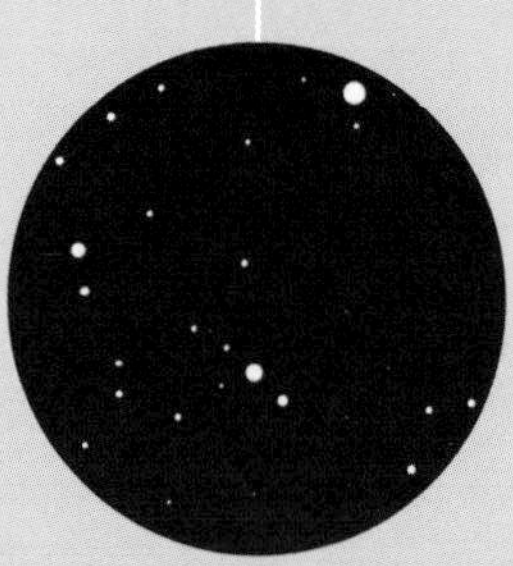

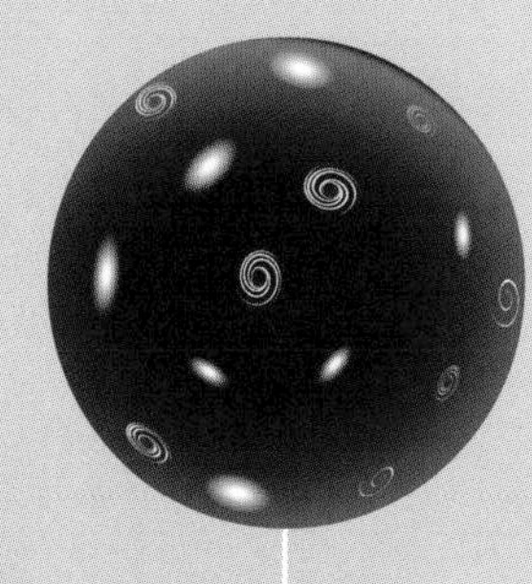

MILKY WAY

Our galaxy, the Milky Way, begins to take on its current form with spiral arms, around 4.5 billion years after it first started to pull together.

SOLAR SYSTEM

The Solar System begins to form as gas and dust pull together to make up the Sun. 100 million years later the Earth forms, with the earliest date for the possible emergence of life just 200 million years after that.

8.7 billion years ago

6 billion years ago

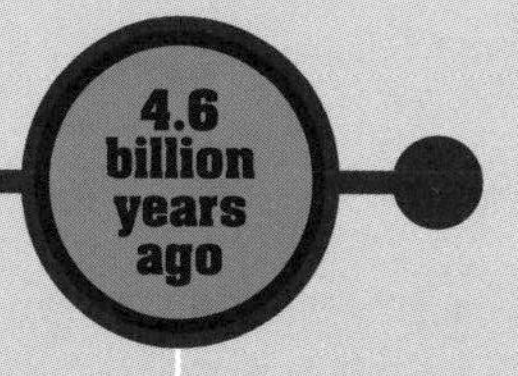

GALAXIES

The first galaxies form as gravitational attraction, possibly assisted by supermassive black holes, pulls collections of stars into larger rotating structures.

DARK ENERGY

The expansion of the universe begins to accelerate as dark energy dominates. Although a tiny effect in any particular locality, over the entire expanse of space, this amounts to around two thirds of the energy content of the universe.

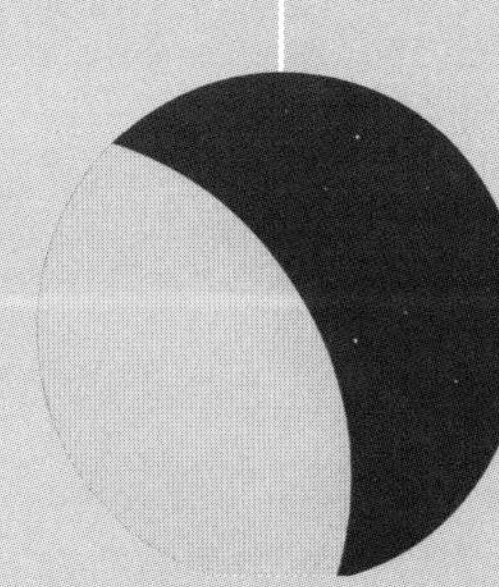

THE VOID

THE MAIN CONCEPT | Before considering motion and relativity we need to have a context. The starting point is totally empty space—a void. It was this that the ancient Greek philosopher Aristotle was referring to when he said that "nature abhors a vacuum." He claimed this because he was opposed to atomic theory. This suggested that matter was made up of tiny fragments that were the smallest it was possible for pieces of matter to be. However, if there were such atoms, then in between them was nothing. A void. Aristotle could not accept that solid matter was riddled with empty space, so dismissed atomic theory, using the argument that if a void did exist, nothing would interact with, say, an arrow in flight, so it would move forever (prefiguring Newton's first law). In a total void there can be no motion or relativity. Motion implies the existence of stuff, whether it's matter or light, to do the moving—to be precise, we need at least one of the particles of the standard model before there can be motion. Relativity is all about how we quantify that motion. It isn't an absolute thing. We need at least one other particle to know how the first particle is moving with respect to the second.

DRILL DOWN | One of the biggest conceptual problems in getting to modern physics was the reality that, on the Earth, we did not have any experience of a vacuum or void until vacuum pumps were invented. Whenever anything moved, it didn't keep moving forever, but rather slowed down to a stop, unless it was constantly being pushed. Ancient Greek philosophers had to invent complex mechanisms to keep arrows flying in the air, as they assumed they would still need something to keep them going after they left the bow, so assumed that the air somehow pulled them along. Similarly, a cart would only keep moving if something pushed or pulled it.

FOCUS | *The requirement not to have a vacuum proved difficult for Aristotle and his contemporaries when it came to space beyond the Moon. This is because they thought this region was eternal and unchanging, so couldn't be made of any of their four normal elements of earth, air, fire, and water. A fifth element, the quintessence was dreamed up to avoid the void.*

ATOMS
Page 20

THE STANDARD MODEL
Page 106

NEWTON'S LAWS
Page 130

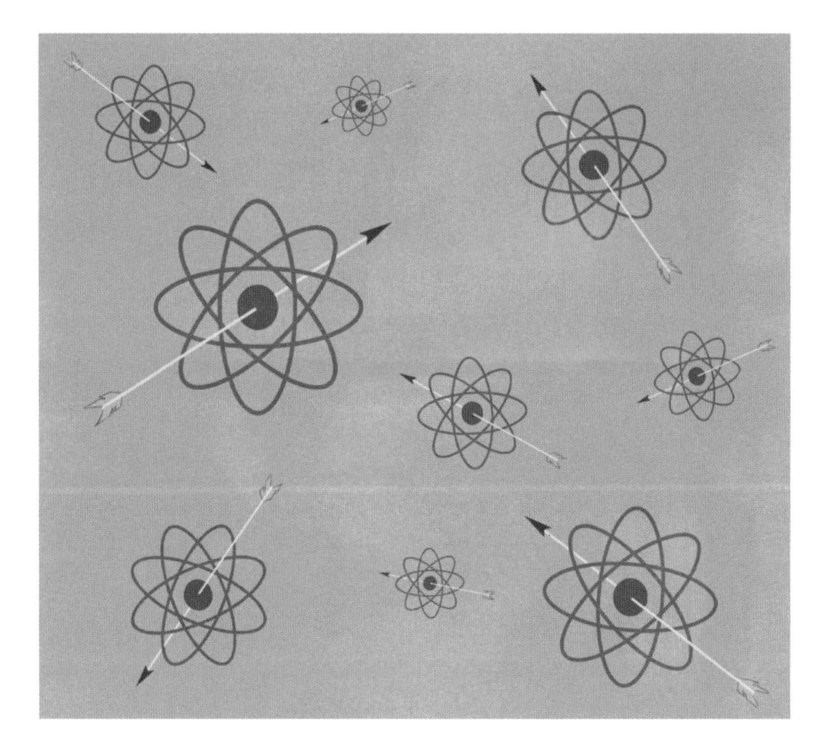

SPEED & VELOCITY

THE MAIN CONCEPT | Once there are objects, they can be in motion. Speed (and its sibling, velocity) gives us a measure of how long it takes an object to reach another object. The existence of speed brings in two other concepts, distance and time. In science, distance is measured in meters and time in seconds, both arbitrary units. When we state a speed, we need to make it clear what that speed is measured with respect to. For example, when we say a car is traveling at 30 meters per second (67 mph), we mean it has that speed with respect to the Earth. If the car is heading straight for a truck, itself traveling at 30 meters per second in the direction of the car, the car's speed with respect to the truck is 60 meters per second. Velocity brings in an extra dimension—quite literally. Velocity is speed in a particular direction. For example, a car with a speed of 30 meters per second could have a velocity of 30 meters per second North. If that were the case, its velocity South is −30 meters per second. A property with a simple numerical value, such as speed, is called a scalar, while one like velocity is a vector.

KINETIC ENERGY
Page 58
MOMENTUM & INERTIA
Page 126
NEWTON'S LAWS
Page 130

DRILL DOWN | The meter was originally 1/10,000,000th of the distance from the North Pole to the equator via Paris. For many years it then became defined as the length of a standard meter rod. However, this was not considered accurate enough, and the meter is now 1/299,792,458th of the distance that light travels in one second. The second gets its name as the second division of an hour (the first being a minute, so called because it's small or "minute"). The idea of there being 60 seconds in a minute comes from the ancient Sumerians, whose number system was based on 60, rather than the base of 10 we now use. In many ways 60 is much easier to handle than 10 in that it is readily divisible by 2, 3, 4, 5, and 6.

FOCUS | *Some physicists argue that it would be far better to measure speed as a fraction of the speed of light. That way, the important universal constant, light speed, becomes simply 1 with no units. Distance would be measured in seconds where 1 second = 299,792,458 meters. The only difficulty is everyday values would be very small: 100 km/h (around 62 mph) would be 0.000000093.*

MOMENTUM & INERTIA

THE MAIN CONCEPT | When an object is in motion it has "oomph" due to its motion, a factor termed its momentum. In scientific terms, the momentum is the mass of the object times its velocity. Momentum, like energy, is conserved in a closed system. When, for example, two objects collide, their total combined momentum remains the same. Because velocity is a vector, having size and direction, so is momentum. So, for instance, a head-on collision of two objects with the same momentum in opposite directions results in zero momentum afterward, as one object will have a momentum of mv and the other –mv; adding them produces zero. This might seem to go wrong when, say, a car hits a wall and stops. Before the collision the car is moving and had momentum. After the collision it has stopped and has no momentum. Where has it gone? It has been transferred to the Earth—the Earth's movement will have been (very) slightly altered. But the mass of the Earth is so much bigger than the car's that the change is unnoticeable. Inertia is sometimes loosely used as an alternative term for momentum, but in physics that word means the tendency of a moving body to keep moving (or a stationary body to stay stationary) unless a force is applied to it.

DRILL DOWN | There is an additional aspect of momentum that arises when things rotate. A separate quantity, angular momentum, is used, and like conventional momentum, angular momentum is conserved. Where ordinary momentum depends on a combination of mass and velocity, angular momentum depends on moment of inertia and speed of rotation. The moment of inertia combines the mass and the distance of the mass from the center of rotation. This is why when a spinning ice skater pulls her arms in, she spins faster. The moment of inertia is reduced because more of the skater's mass is toward the center of rotation, so her speed of rotation increases to conserve angular momentum.

FOCUS | *The conservation of momentum reveals that Hollywood is wrong to show people thrown through the air when hit by a bullet. Rifle bullets have masses of about 0.004 kg and velocities around 1,000 meters per second, giving a momentum of 4 kgm/s. An adult human has a mass of 70 kg. A stopped bullet would move them at 0.06 meters per second—5% of walking pace.*

MASS
Page 22

KINETIC ENERGY
Page 58

NEWTON'S LAWS
Page 130

FORCE & ACCELERATION

THE MAIN CONCEPT | Two essentials to understand motion are force and acceleration. Acceleration is the rate at which velocity changes, and force is the ability to produce acceleration. The relationship between force and acceleration is given by Newton's second law of motion. For a particular object, the acceleration is proportional to the force applied. Inertial mass, according to this law, is simply the force divided by the acceleration. If you think about trying to get something moving, the more mass it has, the harder it is to get it moving. So, for example, it's easy to move a small piece of wood, much harder to move a metal safe with lots of mass. This resistance to getting going (or to stopping once an object is moving) is its inertia. Even when objects are weightless in empty space they still have mass and hence inertia; it still takes force to get them moving. The only difference from being on Earth is that there is no resistance to motion from friction or air resistance, so even a small amount of force will generate some acceleration. When force is applied to an object it is felt as pressure—pressure is simply the force applied divided by the area over which the force is applied.

DRILL DOWN | Pressure is made up of a combination of many small forces because the bodies involved are made up of many small atoms. When you push something with your hand, the force is transmitted from your hand to the object by electromagnetic interaction between the atoms in your hand and the atoms in the object. This is a relatively simple effect as the atoms are acting together, linked in a solid object. When pressure results from a gas—for example the air pressure on the surface of Earth—the interaction is more complex as the gas molecules are independent. The pressure results from the combination of many tiny impacts of the gas molecules on the object feeling the pressure.

MASS
Page 22

MOMENTUM & INERTIA
Page 126

NEWTON'S LAWS
Page 130

FOCUS | *It might seem that because the acceleration of a body is the force applied to the body divided by the body's mass, we would expect more massive objects to accelerate slower as they fall. However, gravitational force goes up with the object's mass and the two cancel out. Ignoring air resistance, everything is accelerated by Earth surface gravity at the same rate.*

NEWTON'S LAWS

THE MAIN CONCEPT | Newton's masterpiece, known as the *Principia* from the first word of its Latin name, introduced two aspects of physics: the laws governing the motion of objects and a mathematical analysis of gravity. The former was shown to involve three simple laws. The first is sometimes called the law of inertia, saying that a moving body will continue moving at the same velocity, and a stationary body will stay stationary, unless a force is applied. The second law describes the scale of the influence of force on movement. It can be stated as force is mass multiplied by acceleration. As momentum is mass times velocity, another way of looking at the second law is that the force applied is equal to the rate at which a body's momentum changes. The third law is usually stated as "every action has an equal and opposite reaction." What this means is that if you apply a force to something it applies an equal and opposite force back. This may sound as if it means that nothing will ever happen, but bear in mind each force is acting on a different body. So, if you push something it will be accelerated—but unless you are well anchored, you will also accelerate in the opposite direction.

DRILL DOWN | Newton's laws of motion can be seen in action when an aircraft is on the runway. Initially, the plane sits stationary, obeying the first law. When the jet engines spin up, they burn fuel rotating a turbine. The air sucked through the engine, plus the jet exhaust, push out of the back of the engine—the third law gives an equal and opposite push forward on the engines, and hence the rest of the plane. As a force is being applied, the second law tells us how the plane accelerates. And the third law makes you feel a push back into your seat in opposition to the seat's forward push.

MOMENTUM & INERTIA
Page 126
FORCE & ACCELERATION
Page 128
SPECIAL RELATIVITY
Page 138

FOCUS | *Newton's third law was misunderstood by the* New York Times *when it attacked rocket pioneer Robert Goddard, saying his rockets would not be able to fly in space because "they need to have something other than a vacuum against which to react." The newspaper assumed a rocket had to have air to push against. In reality, the rocket pushes out its exhaust, and its exhaust pushes back, so it can create thrust anywhere.*

FRICTION

THE MAIN CONCEPT | The ancient Greeks struggled to understand motion because of the ever-present nature of friction. The laws of motion either require us to use a very simple model of the world where everything is perfect and there is no friction, or to ensure that the forces we consider to be acting on a moving body include the less obvious ones, such as those caused by friction and air resistance. Friction is a resistance to movement caused by parts of the moving body rubbing against each other or its surroundings. The result is that energy is lost in heat and the moving body slows down. Sometimes this heat generation is beneficial, such as when we rub our hands together to warm them or strike a match. Friction is partly caused by interactions between tiny irregularities in even the apparently smoothest of surfaces, causing the moving body to drag and slow, but mostly it is due to electrical attraction between atoms. To make matters more complicated, friction comes in two forms—static and dynamic. Static friction (sometimes called stiction) prevents a body that isn't moving from getting started, while dynamic friction occurs during the movement, slowing the body down. Often, static friction takes more to overcome than the dynamic form.

DRILL DOWN | We tend to think of friction as a bad thing, because to engineers, the dragging effect, turning kinetic energy into heat, can be wasteful. This is why we use lubricants to help surfaces move across each other more smoothly. However, friction has positive benefits. Traditional car brakes convert the energy of motion into heat, slowing the car down—though this wasteful process is bettered in electrical cars, where the energy is used to recharge the battery. And an entirely frictionless world would make it impossible to do anything. Think of having to live on an ice rink, but with even more slippery surfaces. We couldn't walk or pick anything up without friction.

BONDING
Page 24

KINETIC ENERGY
Page 58

NEWTON'S LAWS
Page 130

FOCUS | *The ultimate example of static friction at work is in the feet of geckos, which can climb up smooth walls or even glass. On a gecko's toes, millions of tiny hair-like protuberances called setae ensure there is far more contact with the surface than would normally be the case. It's the tiny electrical attractions between atoms that cause them to stick.*

FLUID DYNAMICS

THE MAIN CONCEPT | When we learn about the laws of motion at school, there is rarely mention of fluids, which means that the way that gases and liquids flow is pretty much absent from nonscientists' concept of physics. Yet it's hugely important in the real world, whether we are trying to understand major Earth systems, such as the way movement of the Earth's molten iron core produces magnetism or how ocean currents influence climate, or to deal with everyday essentials, such as getting the right-sized pipe for a required flow of water or gas. There's a reason we aren't taught this kind of thing in school—the mathematics involved is messy. This is because of the complexity of interaction between different parts of a body of fluid, resulting in movement that can easily become chaotic. Successfully modeling fluid flow is, in principle, just a matter of applying Newton's laws of motion to small segments of the fluid to see how factors such as density, temperature, velocity, and pressure influence the overall movement. There are equations to deal with this, the Navier-Stokes equations, but solving them for any real system is near impossible. Usually a solution is approximated and refined using a computer. Once flow becomes turbulent the calculations become pretty much impossible to carry through and significant approximations are required.

DRILL DOWN | The chaotic aspect of fluid flow is why weather forecasting, dealing with the flow of vast bodies of air, is difficult. It is impossible to meaningfully forecast more than ten days ahead. The mathematics of chaotic systems—those where a small change in initial conditions make a huge difference in outcome—began with an early weather forecasting computer program. American meteorologist Edward Lorenz wanted to rerun a forecast, so he inputted the data from a printout. The result was totally different. This happened because the program printed out fewer decimal places than the computer used. A tiny change in the input resulted in a transformed forecast—because of the chaos inherent in complex fluid dynamics.

FOCUS | *Understanding fluid dynamics is crucial in aircraft design. Some modern airliners have winglets—relatively small extensions from the tips of the wings. This is because the tips of the wings produce turbulent flow as they move through the air, causing vortices to form in the air, which result in drag. The winglets cut through the spinning air, reducing the drag effect.*

SOLIDS & LIQUIDS
Page 26
GASES & PLASMAS
Page 28
NEWTON'S LAWS
Page 130

GALILEAN RELATIVITY

THE MAIN CONCEPT | We associate "relativity" with Albert Einstein, but Galileo was the first to explore the implications of this essential aspect of physics. Relativity really means that, in dealing with matter in motion, we need to consider what the matter is moving with respect to. Because the Earth is so large, it's easy to think of the Earth's surface as something universal and fixed, but, of course, it's not. If you're at home, reading this book sitting in a chair, the easy assumption is that you are not moving. But Galilean relativity says that what you really mean is that you are not moving relative to the seat—and hence the Earth. Compare yourself with a point out in space and you are hurtling through space at thousands of miles an hour as the Earth rotates and passes around its orbit. All movement is relative and we always need to establish the "frame of reference," that is, what we are measuring that movement relative to. It is possible that you are reading this book on a train or plane. In that case, you would probably say that you are, indeed, moving—because of the dominance of the Earth's surface as a frame of reference. But relative to the train or plane you are stationary.

TUNNELING
Page 100
THE VOID
Page 122
MOMENTUM & INERTIA
Page 126

DRILL DOWN | Aware of the importance of relativity to motion, Galileo imagined traveling in a smooth boat at constant speed. The boat had no windows, and was so smooth that it was impossible to feel any sensation of movement. Galileo said that no physical equipment within the boat would be able to detect whether or not it was moving. You could observe pendulums swinging, objects falling or rolling down inclined planes and their motion would be exactly as you would expect in the room of a house. This was because, as far as the experiments were concerned, the boat was not moving. Only if the vessel accelerated would it be possible to detect an effect.

FOCUS | *Galileo is said to have demonstrated relativity in a fast boat on Lake Piediluco. He borrowed a key from a friend named Stelluti and threw it straight up in the air. Other friends had to restrain Stelluti from jumping in the water, assuming the key would be left behind—but it fell back in Galileo's lap. For the key, the boat wasn't moving.*

SPECIAL RELATIVITY

THE MAIN CONCEPT | In 1905, Albert Einstein, at the time still a clerk in the Swiss patent office in Bern, had four major papers published, one of which covered the special theory of relativity, and another a consequence of this theory, namely that $E=mc^2$. The special theory combines Newton's laws of motion with the requirement from Maxwell's electromagnetic theory that light moves at a fixed speed in any particular medium. The implications of the special theory are remarkable. As a result of movement, events that appear to be simultaneous cease to be so. Relative to a point that's not moving at the same velocity, time slows on a moving body, mass increases, and the dimension of the body in the direction of movement gets smaller. We do not generally notice this, as the effect only becomes significant when the body is moving extremely quickly. The effect has been tested many times, most notably in the aspect of slowing time, which provides a mechanism for time travel into the future. This was first tested by flying a highly accurate atomic clock around the world and comparing it to an identical clock that remained on the ground. As a result of the special theory, the previously separate concepts of space and time have to be considered as a unified spacetime.

DRILL DOWN | The time on a moving body slows down, so if it travels away from the Earth and then returns, time will have moved quicker on Earth and the body will have moved into the Earth's future. A thought experiment called the twins paradox sends one twin off in a very fast spaceship and leaves the other on Earth. When the traveling twin returns, according to her it might be 2050, while on Earth it's 2070. The space-traveling twin has traveled 20 years into the future and is 20 years younger than her Earthbound sister. The difference between the pair is that only the twin in the ship undergoes the powerful acceleration and deceleration that results in her time travel.

SPEED OF LIGHT
Page 44

ELECTROMAGNETISM
Page 88

NEWTON'S LAWS
Page 130

FOCUS | *Unlike quantum physics, special relativity has few everyday applications, but is involved in the working of GPS satellites providing satellite navigation. Because the satellites move with respect to the Earth, special relativity causes time onboard to run slightly slowly. As GPS works by broadcasting an accurate time signal, the timing has to be corrected for this effect (and for general relativity, which also influences time).*

GRAVITY

THE MAIN CONCEPT | According to Newton's first law, a planet flying through space should continue to travel in a straight line unless a force acts on it. Newton realized that a force of gravity, acting to pull a passing object inward, would result in the kind of orbits we see when the Moon travels around the Earth, and the Earth travels around the Sun. He calculated that such attraction would have the same magnitude as we observe when something is dropped on the Earth—an inward acceleration, which at any particular height above the Earth's surface would be the same for falling objects and bodies in orbit. Newton produced detailed mathematical arguments, showing that the orbits observed by astronomers would result from an inverse square law, where the force of attraction was proportional to one over the square of the distance between the bodies involved. He did not explicitly state it, but his workings showed that the force of gravity was equal to Gm_1m_2/r^2, where G is a constant value, m_1 and m_2 are the masses of the bodies, and r is the distance between them. This simple equation was all that was necessary to plot the motion of the planets and, hundreds of years later, to guide a rocket to the Moon.

DRILL DOWN | Newton's representation of gravity as an attractive force between two bodies replaced the concept that had held sway since the time of the ancient Greeks. This was based on the idea that matter was made up of four elements. Of these, two (earth and water) had gravity, a tendency to move to the center of the universe, and the other two (air and fire) had levity, a tendency to move away from the center of the universe. As the Copernican model of the universe, which had the Sun at the center rather than the Earth, was taking hold by Newton's time, the old concepts of gravity and levity could no longer apply.

MASS
Page 22

POTENTIAL ENERGY
Page 60

NEWTON'S LAWS
Page 130

FOCUS | *When Newton published his work on gravity it was mocked by some of his contemporaries because it involved a mysterious attractive force, say between the Earth and the Moon. The word "attraction" seems perfectly normal now, but when Newton used it, it was only applied to a physical attraction between people. He seemed to be suggesting the Moon found the Earth beautiful.*

GENERAL THEORY OF RELATIVITY

THE MAIN CONCEPT | The special theory of relativity was a significant piece of work in its own right, but it did not cover all kinds of accelerating body. Einstein had a revelation when he thought that bodies in free fall don't feel their own weight, realizing that this meant that there was an equivalence between gravity and acceleration. If you were in a spaceship sitting on the Earth, feeling the Earth's gravity, you could not distinguish between that and an accelerating ship producing a force of 1 G. A simple thought experiment shows how this implies that gravity warps spacetime. Imagine a beam of light crossing the interior of a spaceship. If the ship accelerates, the beam will bend, as the ship accelerates during the time it takes light to cross it. But if gravity and acceleration are equivalent, gravity would also cause the beam to bend. The mathematics of general relativity were beyond Einstein, and it was only with help from others that he was able to put together formulae involving the geometry of curved space to produce his equations for the general theory of relativity. These combine four factors that result in the exact behavior of gravity we observe, which is subtly different from the one described by Newton's mathematics, and more importantly the theory gives a reason for the gravitational pull.

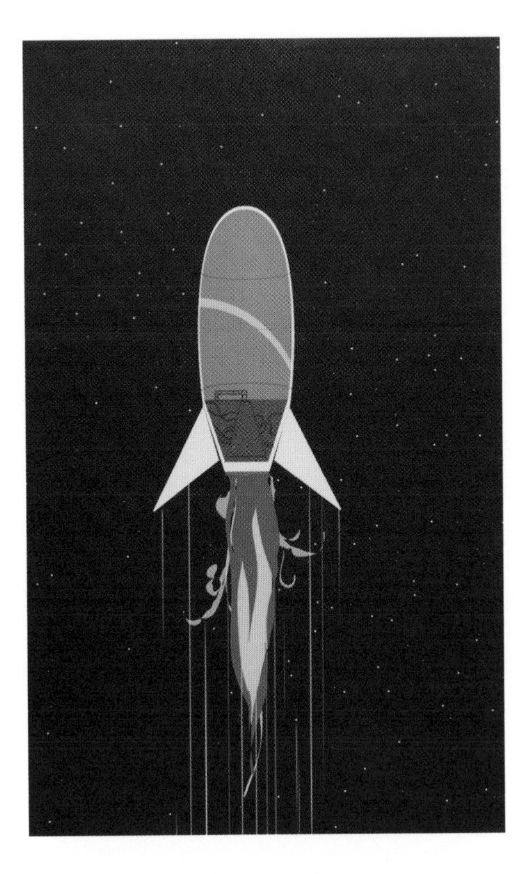

FORCE & ACCELERATION
Page 128
SPECIAL RELATIVITY
Page 138
GRAVITY
Page 140

DRILL DOWN | The idea that a warp in space could produce an orbit is simple. Imagine a taut rubber sheet is space and a straight line drawn on it is the path of a planet. If a bowling ball is placed on the sheet, warping it like a mass warps space, the straight line becomes a curve around the ball. However, this doesn't explain falling objects. They would only roll down the sheet toward the ball if gravity applied, which is a circular argument as we are trying to produce gravity. Things fall because mass doesn't just warp space, but spacetime—in effect, a change in temporal position becomes a change in spatial position.

FOCUS | *When someone doesn't feel their weight on the International Space Station, it's not zero gravity—in fact at that altitude, gravity is around 90 percent of the pull at the surface of Earth. Instead, he or she is in free fall toward the Earth; their acceleration cancels out gravity. Luckily the ISS is both falling and moving sideways (the definition of an orbit), so it misses the Earth.*

BLACK HOLES

THE MAIN CONCEPT | Within months of Einstein publishing his general theory of relativity, another German physicist, Karl Schwarzschild, had solved the equations for a very simple case of a non-rotating sphere. He discovered that if such a body were dense enough there would be a distance away from its center, later called the Schwarzschild radius, within which spacetime would be so warped that nothing could escape, not even light. A sphere with this radius is known as the black hole's event horizon. Such a body had been suggested for different reasons by English natural philosopher John Michell in the eighteenth century, but Schwarzschild's concept emerged from the nature of gravity. This phenomenon, hypothetical at the time, was named black hole by an unknown contributor and then popularized by American physicist John Wheeler. Wheeler, along with Stephen Hawking, was among the leaders in considering the theories of black holes and other gravitational oddities, such as wormholes. By the end of the twentieth century there was strong evidence that black holes did exist. Though they can not be seen, their impact on their surroundings is detectable. Supermassive black holes appear to form the central hubs of most galaxies, while collisions of black holes have recently been detected using gravitational waves, predicted by Einstein in 1916.

DRILL DOWN | When black holes collide they produce ripples in spacetime, which travel outward at the speed of light. Any movement of massive objects should do this, but usually the effect is unnoticeably small. Einstein predicted gravitational waves, but also that they would never be discovered as the effects would be so tiny. For over 50 years, scientists searched for gravitational waves until 2015 when the LIGO observatory, a pair of massive US detectors, picked up ripples from colliding black holes. It's not surprising it took so long. To find a wave, the devices had to detect the movement of a mirror by 100th of the diameter of the nucleus of an atom.

FOCUS | *Thinking about escape velocity led John Michell to predict the existence of black holes in 1783. Escape velocity is the speed needed to get away from the surface of a body. Anything thrown from Earth at 11.19 kilometers/second (6.95 miles/second) will keep going, but if slower it will fall back. Michell argued that a dense star's escape velocity could be greater than light speed.*

GRAVITY
Page 140
GENERAL THEORY OF RELATIVITY
Page 142
MODELING THE UNIVERSE
Page 146

MODELING THE UNIVERSE

THE MAIN CONCEPT | Although the first use of a solution of Einstein's general relativity equations was a simple non-rotating sphere, theorists were soon ready to take on a bigger challenge—the whole universe. Clearly it was no more possible to model every star in the universe than it is possible to plot the movement of every atom in a gas, but Einstein was first with a simple model in 1917. However, he realized there was a problem: if, as was assumed at the time, the universe was neither contracting nor expanding, eventually it would collapse as all the matter scattered through the universe was attracted to the rest. To keep things stable, Einstein added a small extra part to his equations known as the cosmological constant. Once it was discovered in the 1920s that the universe was indeed expanding, an alternative model developed by the Russian cosmologist Alexander Friedmann made the constant unnecessary. Einstein referred to it as his "greatest mistake." Work by another cosmologist, Belgian priest Georges Lemaître, led to the core of the "big bang" model of the expanding universe still used today. However, Einstein's constant turned out not be a mistake after all. It was reinstated in the 1990s when it was discovered that the universe was expanding faster than predicted, powered by an unknown mechanism called "dark energy."

DRILL DOWN | The expansion of the universe was deduced from data produced by American astronomer Edwin Hubble. It was Hubble who had first calculated distances to other galaxies and showed that they were not, as many thought, simply parts of the Milky Way. However, he also noted that the farther away the galaxies were, the more they were red-shifted. This is an effect of light where, when something is moving away, the wavelengths of the light increase, moving the color toward the red end of the spectrum. It was Hubble's data—using the discovery of American astronomer Henrietta Swan Leavitt that variable stars could act as distance measures known as standard candles—that cemented the idea of an expanding universe and hence one that could begin in a big bang.

GRAVITY
Page 140
GENERAL THEORY OF RELATIVITY
Page 142
BLACK HOLES
Page 144

FOCUS | *The idea of the universe collapsing due to the attraction between stars was nothing new. Newton had considered it and decided that this implied the universe was infinite and uniform, so no stars would feel more pull inward than outward. This would be potentially unstable if any star moved slightly, so Newton assigned God the task of poking stars back in place.*

GLOSSARY

ANTIMATTER—Alternative form of matter where particles have the opposite charge and differ in other quantum properties from ordinary matter. When matter and antimatter particles collide, they convert into pure energy.

ATOM—The smallest part a chemical element can be divided into. It was originally thought that atoms were the smallest fundamental particles (the name means "uncuttable"), but we now know that atoms have an internal structure.

CHARGE—A property of a particle. Usually refers to electrical charge, which has opposing "flavors" of positive and negative. The basic unit of charge is that of the electron (–1) or proton (+1), however this was established before the discovery of quarks, which have charges that are multiples of $\frac{1}{3}$ (or $-\frac{1}{3}$). There are other "charge" properties, such as the color charge that applies to quarks.

DARK ENERGY—An unknown energy source causing the universe to expand at an accelerating rate. If we add together all the matter and energy in the universe, dark energy accounts for around 68 percent of the total content of the universe.

DARK MATTER—A hypothetical type of matter that interacts through gravity but not electromagnetically, so it is invisible and passes through ordinary matter. Devised to account for the behavior of large bodies such as galaxies, which act as if they had a lot more matter in them than is believed to be there. If it exists, there is around five times as much dark matter as ordinary matter, accounting for around 27 percent of the content of the universe. However, some physicists believe dark matter does not exist and the effect is caused by a variation in the force of gravity on the scale of galaxies.

DENSITY—The amount of mass in a given volume of a substance. The bigger the density, the more mass the same size object will have.

ELECTROMAGNETISM—The field of physics dealing with electricity and magnetism. In the nineteenth century the two phenomena, which had been known since ancient times, were discovered to be aspects of the combined electromagnetism.

ELECTRON—A fundamental particle of matter and the first component of the atom to be discovered. The distribution of electrons around an atom is responsible for an element's chemical properties and a flow of electrons makes up an electrical current.

ELEMENT—All matter is made up of a small number of chemical elements. Around 94 exist in nature, while a few more can be made artificially. Each element is made up of identical atoms with the same configuration of protons and electrons, but can have differing numbers of neutrons, producing isotopes, which are chemically identical atoms but with different masses.

ENERGY—A natural phenomenon that makes things happen (does work) and can be present in the form of heat, movement (kinetic energy), etc. Energy and matter are interchangeable and the total amount of energy and matter in an isolated space always remains the same.

ENTROPY—A measure of the amount of disorder in a system. The greater the disorder, the higher the entropy. The amount of entropy in an isolated system will stay the same or increase, though energy from outside can be used to decrease the level of entropy.

FIELD—A physical phenomenon that has a value for any point in time and space, which can result in apparent influence at a distance—such as the electromagnetic field around a magnet or the gravitational field around a planet.

FISSION, NUCLEAR—The nucleus of a heavy atom can split into two or more parts, releasing energy. This splitting can occur spontaneously or as a result of collision with another particle. Such splitting is known as nuclear fission, and is our current main source of nuclear energy, as well as being used in nuclear weapons.

FREQUENCY—When a phenomenon occurs regularly, for example a wave, its frequency is the number of times the phenomenon goes through a cycle and returns to its starting condition in one second. Frequency is measured in hertz (Hz) so, for example, a 200 Hz sound wave undergoes 200 complete cycles in a second.

FUSION, NUCLEAR—The nuclei of two or more lighter atoms can join together to form a heavier atom. In the process, energy is given off. This is the power source of stars and is used in experimental nuclear reactors, as it produces clean nuclear energy. Nuclear fusion also occurs in "thermonuclear" weapons or hydrogen bombs.

GLUON—A massless particle that "glues" together the quark particles that make up protons, neutrons, and some other compound particles.

HEAT ENGINE—A mechanism for turning heat into mechanical work. Originally applied to a steam engine, but internal combustion engines and the turbines used in power stations are all heat engines.

INERTIA The tendency of a moving body to keep moving at the same speed unless something acts on it to slow it down or speed it up.

INTERFERENCE—A wave effect in which two waves cross each other. If the oscillations of the waves are heading in the same direction, they add together to produce a bigger wave. Similarly, if they are oscillating in opposite directions, they cancel each other out. This is known as interference.

ION—An atom that has lost one or more electrons, becoming positively charged, or has gained one or more electrons, becoming negatively charged.

MASS—A measure of the amount of matter in an object. Mass controls how much acceleration the object gains under a particular force or the force of gravity it feels.

MATTER—Stuff. The material that makes up solid objects, liquids, and gasses. Matter is composed of atoms, which may be joined together to form molecules.

MECHANICS—The area of physics that deals with moving objects and how their motion is altered when forces are applied to them.

MOLECULE—Two or more atoms joined together to form the smallest component of many substances. In some cases the atoms are of the same element, but they often contain a combination of elements.

MOMENTUM—The mass of an object multiplied by its velocity to give a value akin to the quantity of motion.

NUCLEUS (plural **NUCLEI**)—The small central part of an atom containing most of its mass in proton and neutron particles.

NEUTRINO—A very light particle, typically produced in nuclear reactions, and which has very little interaction with matter particles.

NEUTRON—One of the two components of the nucleus of an atom, which with protons make up the bulk of the mass of an atom. Neutron particles have no electrical charge.

PARTICLE—An extremely small component of nature, making up matter and other natural phenomena such as light. Some particles, such as electrons and photons, are thought to be fundamental—not having components of their own—while others, such as protons and neutrons, are made up of smaller particles.

PHOTON—A particle of light. For many years, light was thought to be a wave, but quantum theory showed that it could behave both as a particle and as a wave, though not both simultaneously.

PLASMA—A gas made up of ions rather than atoms or molecules. Often called the fourth state of matter, plasmas are good electrical conductors, and make up the majority of the matter in the universe, as stars are primarily plasma.

POSITRON—The antiparticle of an electron. This antimatter particle is like an electron but has a positive rather than a negative charge.

PROBABILITY—The mathematical chance of something happening. Probability theory is important in quantum physics, where particles do not, for example, usually have defined locations but exist as probabilities of being in a range of locations until they interact with another particle.

PROTON—One of the two components of the nucleus of an atom, which with neutrons make up the bulk of the mass of an atom. Proton particles are electrically positive, with the same size (but opposite) charge as an electron.

QUANTUM—A quantity of something. Quantum physics is based on the idea that phenomena usually come in particles with a minimum size, rather than being of continuously variable size.

QUARK—Fundamental matter particle which come in six "flavors." Combinations of quarks make up particles such as protons and neutrons.

SYSTEM—In the context of physics, this is the entire contents of a particular volume of space. An isolated system does not interact with matter or energy outside it, while an open system can do so.

THERMAL EQUILIBRIUM—When the heat passing between different parts of a system balances out, so that there is no net flow of heat in any direction.

VACUUM—A volume of space which contains no (or very small amounts of) matter.

VOLUME—The amount of three-dimensional space that an object occupies.

WAVE—A regularly repeating oscillation in a material or field, which can carry energy.

WAVELENGTH—The distance between repetitions in a wave.

WEIGHT—The force due to gravity on an object of a particular mass. Your weight would be reduced to a sixth of its usual value if you were on the surface of the Moon, but your mass would be unchanged.

WORK—The energy involved when a force makes a change happen.

FURTHER READING

BOOKS

General Physics

Ananthaswamy, A. *The Edge of Physics*. Boston: Houghton Mifflin Harcourt, 2010.

A tour of the most weird and wonderful aspects of physics.

Baggott, J. *Higgs*. Oxford: Oxford University Press, 2012.

Fascinating story of the discovery of the Higgs boson, what it is, and why it matters.

Baggott, J. *Mass*. Oxford: Oxford University Press, 2017.

Detailed consideration of the nature of matter, and particularly its property of mass, bringing in everything from quantum theory to general relativity.

Carroll, S. *From Eternity to Here*. New York: Dutton, 2010.

Not an easy read, but the book *A Brief History of Time* should have been in its exploration of the nature of time and entropy.

Clegg, B. *Light Years*. London: Icon Books, 2015.

The story of humanity's relationship with light and the physics behind it.

Couper H. and Henbest, N. *The Story of Astronomy*. London: Cassell, 2012.

The history of the oldest of the physical sciences, told extremely well.

Feynman, R. P. *Surely You're Joking, Mr. Feynman!* New York: W. W. Norton & Company, 1985.

Autobiographical stories of the development of physics including the Manhattan Project from the twentieth century's most charismatic physicist.

Gregory, B. *Inventing Reality*. New York: Wiley, 1990.

Not an easy read, but the definitive book on the nature of physics and its relationship with the actual world.

Kaku, M. *Physics of the Impossible*. London: Allen Lane, 2008.

The best explanation of the physics that could make science-fiction technology possible.

Muller, R. A. *Physics for Future Presidents*. New York: W. W. Norton & Company, 2008.

A fascinatingly different approach to physics, looking at what a future US president ought to know.

Rovelli, C. *Seven Brief Lessons on Physics*. New York: Riverhead Books, 2016.

Seven short essays on some of the most striking aspects of modern physics, including Rovelli's specialty, loop quantum gravity.

Smolin, L. *Time Reborn*. Boston: Houghton Mifflin Harcourt, 2013.

Many physicists like to say that time doesn't exist—this fascinating title explains why, but also shows how the idea of time being an illusion is itself mistaken.

Tyson, N. D. *Astrophysics for People in a Hurry*. New York: W. W. Norton & Company, 2017.

A very approachable introduction to astrophysics from the TV astronomer.

Weinberg, S. *To Explain the World*. New York: HarperCollins, 2015.

The Nobel Prize-winning physicist puts physics into the context of the invention of science.

Quantum Physics

Ball, P. *Beyond Weird*. Chicago: University of Chicago Press, 2018.

An exploration of the different interpretations of quantum physics.

Clegg, B. *Crash Course: Quantum Physics*. New York: Barnes & Noble/London: Ivy Press, 2019.

A companion to this book, specializing in quantum physics.

Clegg, B. *The God Effect*. New York: St. Martin's Griffin, 2009.

An in-depth look at quantum entanglement, the strangest aspect of physics that enables particles to communicate instantly at any distance.

Clegg, B. *The Quantum Age*. London: Icon Books, 2015.

Detailed coverage of the applications of quantum physics from electronics to MRI scanners, including the physics behind them.

Close, F. *Neutrino*. Oxford: Oxford University Press, 2010.

The story of the hunt for the elusive quantum particle that is at the heart of nuclear reactions.

Farmelo, G. *The Strangest Man*. London: Faber & Faber, 2009.

Biography of the quantum physicist Paul Dirac, giving details of the history of quantum theory.

Feynman, R. P. *QED*. Princeton: Princeton University Press, 1985.

An introduction by physicist Feynman to quantum electrodynamics, the quantum physics of light and matter, the subject that won him the Nobel Prize.

Relativity

Chown, M. *The Ascent of Gravity*. London: Weidenfeld & Nicolson, 2017.

High-level introduction to the nature of gravity and the general theory of relativity.

Clegg, B. *How to Build a Time Machine*. New York: St. Martin's Griffin, 2011.

How both the special and general theories of relativity make time travel possible.

Clegg, B. *The Reality Frame*. London: Icon Books, 2017.

Explores Galilean, special, and general theories of relativity in the wider context of humanity's relationship to the universe.

Isaacson, W. *Einstein: His Life and Universe*. New York: Simon & Schuster, 2008.

There are many scientific biographies of the best-known modern physicists but few rival this.

Scarf, C. *Gravity's Engines*. London: Allen Lane, 2012.

Fascinating description of the nature of black holes and how they may have shaped our galaxies.

WEB SITES

APS Physics

www.aps.org

Web site of the American Physical Society

New Scientist

www.newscientist.com/subject/physics

Physics section of the *New Scientist* site

Physics.org

www.physics.org

Physics web hub from the Institute of Physics

Physics World

www.physicsworld.com

Web site of *Physics World* magazine

Popular Science book reviews

www.popularscience.co.uk

Book review site for popular science books with over 200 physics titles

Scientific American Physics

www.scientificamerican.com/physics

Physics section of the *Scientific American* site

INDEX

ABOUT THE AUTHOR

Brian Clegg With MAs in Natural Sciences (specializing in experimental physics) from Cambridge University and Operational Research from Lancaster University, Brian Clegg (www. brianclegg.net) worked at British Airways for 17 years before setting up his own creativity training company. He has been a full-time science writer for 15 years with over 30 titles published from *A Brief History of Infinity* (2003) to *The Quantum Age* (2015), and most recently *Professor Maxwell's Duplicitous Demon* (2019). He has also written for a range of publications from *The Wall Street Journal* to *BBC Focus* and *Playboy* magazines. He lives in Wiltshire, England.

ACKNOWLEDGMENTS

Author Acknowledgments

For Gillian, Chelsea and Rebecca

Thanks to all at Quarto, notably Caroline Earle and Tom Kitch, and to the University of Cambridge for converting me from wanting to do chemistry to physics.

Picture credits

The publisher would like to thank the following for permission to reproduce copyright material:

Alamy 17TR.

Clipart 50BL, 52BL, 52 TR.

Getty Images/Bettmann: 16TR, 85BL, 85TR; Hulton Archive/Stringer: 51TR; Photo12: 17BL; Science & Society Picture Library: 51BL.

Ivy Press/Andrea Ucini: 1, 3, 7, 9 14, 15, 23, 27, 29, 31, 33, 35, 37, 39, 41, 45, 48, 49, 52TL, 53TR, 53BL, 53BR, .55, 57, 59, 61, 63, 65, 67, 69, 71, 75, 75, 77, 79, 82, 83, 86TL, 86TR, 86BL, 86BR, 87TL, 87TR, 89, 91, 93, 95, 97, 101, 103, 107, 109, 111, 113, 116, 117, 120TR, 120BR, 121TR, 121BR, 123, 125, 127, 129, 131, 133, 137, 139, 141, 143, 145, 147.

Ivy Press/Nick Rowland: 18TR, 19TC, 19BR, 21, 25, 43, 53TL, 87BL, 87BR, 99, 105, 135.

Library of Congress, Washington D.C. 16BL, 118BL, 119BL.

Shutterstock/Alex Mit: 121BL; Designua: 19BL, 121TL; Georgios Kollidas: 118TR; IgorZh: 120TL; Morphart Creation: 52BR; Sergey Nivens: 120BL; Twocoms: 119TR.

Wellcome Collection/CC BY 4.0: 18BC, 50TR, 84BL, 84TR.

All reasonable efforts have been made to trace copyright holders and to obtain their permission for the use of copyright material. The publisher apologizes for any errors or omissions in the list above and will gratefully incorporate any corrections in future reprints if notified.